# WEIGHING THE WORLD

# WEIGHING
## THE
# WORLD

*The Quest to Measure the Earth*

EDWIN DANSON

OXFORD
UNIVERSITY PRESS

2006

# OXFORD
## UNIVERSITY PRESS

Oxford University Press, Inc., publishes works that further
Oxford University's objective of excellence
in research, scholarship, and education.

Auckland   Cape Town   Dar es Salaam   Hong Kong   Karachi
Kuala Lumpur   Madrid   Melbourne   Mexico City   Nairobi
New Delhi   Shanghai   Taipei   Toronto

With offices in
Argentina   Austria   Brazil   Chile   Czech Republic   France   Greece
Guatemala   Hungary   Italy   Japan   Poland   Portugal   Singapore
South Korea   Switzerland   Thailand   Turkey   Ukraine   Vietnam

Copyright © 2006 by Oxford University Press, Inc.

Published by Oxford University Press, Inc.
198 Madison Avenue, New York, New York 10016

www.oup.com

Library of Congress Cataloging-in-Publication Data
Danson, Edwin, 1948–
Weighing the world : the quest to measure the Earth / Edwin Danson.
p. cm.
Includes bibliographical references and index.
ISBN-13: 978-0-19-518169-2
ISBN-10: 0-19-518169-7
1. Earth—Figure—Measurement.   2. Science—Europe—History—18th century.   3. Industrial
revolution—Europe, History.   I. Title.
QB334.D36 2005
526'.1—dc22      2004066284

1   3   5   7   9   8   6   4   2

Printed in the United States of America
on acid-free paper

For he hath weighed the world in the balance.
By measure hath he measured the times,
And by number hath he numbered the times.

Second Book of Esdras, ch. IV, vs. 36–37


He had bought a large map representing the sea,
Without the least vestige of land:
And the crew were much pleased when they found it to be
A map they could all understand.

Lewis Carroll, "The Hunting of the Snark"

# *Preface*

IT WAS AN AGE OF REASON; it was an age of enlightenment. It was an age of philosophical and scientific revolution, a fleeting period in history sandwiched between two momentous political revolutions: the bloodless English revolution of 1688 and the very bloody French Revolution of 1789.

The golden age began with the sowing of the mechanical seeds that grew to become the Industrial Revolution. The opening decade of the eighteenth century saw the world's first wheezing steam plant, and by 1712 Thomas Newcomen's steam-powered pumping engine was sucking prodigious amounts of water from a coal mine in Derbyshire. When the century began, manufacturing throughout the Western world was a cottage industry; 80 years later Sir Richard Arkwright had set in motion the "factory revolution" and was employing over 5,000 workers in his dark, but not yet "satanic," textile mills.

It was the age of Swift and Johnson, of political radicals such as Wilkes, and of philosophers like Rousseau and Voltaire. Mozart was born and died in this century, the genius of Beethoven flourished, and Handel wrote music sublime to entertain kings. And there were the poets, such as Pope, Goldsmith, and Cowper, who captured the essence of change that was sweeping across Europe. "Whereas in France the hurricane of revolution swept the country," reflected Engels in his socialist review of the eighteenth century, "there passed through England a quieter, but no less powerful upheaval. Steam and the new mechanical tools changed mill-working into modern heavy industry, thereby revolutionising the whole basis of middle class society. The sleepy evolution of the period of manufacture was turned into a veritable storm . . . of production."[1]

It was a turbulent age, a time of nearly continuous conflict by armies and navies increasingly furnished with scientific wonders to improve the means of waging war in ever more terrible, more destructive, and more devastating ways. The first decade of the century saw the union of Scotland with England and the beginnings of worldwide political unrest. Across the Atlantic Ocean, Britain's colonies in America grew and flourished, competing in affluence and power with their French and Spanish neighbors. The settlers of the eastern seaboard were already beginning to wonder why they needed the old country at all.

The great European conflict of 1755, known as the Seven Years' War, was kindled in the backwoods of America. It was the first world war and was fought across North America, up and down the coasts of India, and amid the tropical islands of the East Indies. It was fought across Europe and on the high seas, along the Africa coasts, and among the sugar islands of the West Indies. The victor's spoils would be the foundation of empire.

The great American sage Benjamin Franklin and other enlightened men on both sides of the Atlantic suddenly recognized a unique instant in world history, an opportunity for a "wider British Empire,"[2] one that offered horizons of influence and commerce undreamed of. But it was not to be; the opportunity was cast away for the selfish interests and narrow perspectives of a wealthy few.

Not yet an empire, not yet a nation so divided by the evils of the Industrial Revolution as to be without a humanist core, Britain had nevertheless taken the first steps toward a new and disturbing form of class segregation. For 700 years, its common people had lived the legacy of the Norman invasion, dominated by a Norman-French aristocracy that was wealthy almost beyond reason. Now the inventive and exploitative merchant classes, with their "new money," were adding new layers to the social hierarchy in fulfillment of their own special labor needs. The eighteenth century, then, in its later years saw the rise of a disadvantaged proletariat and a hardening of the ubiquitous British class system, with its rigid, illogical, and slightly ridiculous (but ever entertaining) social divides.

Across Europe, the art of thinking was undergoing a revolution equally as profound as that in artillery or politics. Freethinking intellectuals, the first of their genre, were employing reason and science to challenge traditional values, question conventional ideas, and contest long-held beliefs. In Britain, writers like Jonathan Swift were using "the artillery of words" to electrify the newly emergent, affluent, and literate middle classes and revolutionize the accepted wisdom of a thousand years. The eighteenth century, unlike its successor, represented an almost unique period in history, when bright young boys of the humblest birth might become rich merchants, great military men, or famous scientists.

For scientific advance, the eighteenth century saw the culmination of the slow transition of the ancient "sciences" such as alchemy and astrology into the "new" sciences of chemistry, physics, and astronomy. Issac Newton, in the previous century, wrote significantly more on alchemy and the harmony of nature than on his gravity theories. Astrology was still a respectable pursuit, and many of the great astronomers of the age were astrologers. With the thirst for learning, so new sciences evolved, particularly that of geography, geology, and mineralogy as mankind's curiosity and the demands for raw minerals to fuel the Industrial Revolution awakened latent intellects.

Before Engels's veritable storm could unleash its full fury, the obstacles to its passage, and to that of trade itself, had to undergo their own quiet revolution. Commerce was hampered by a lack of good roads; communications throughout the developing world were much as they had been in the Middle Ages. The economies of the trading nations of Europe were increasingly dependent on foreign trade to fuel their growth and pay their armies. Cargoes were ever at the mercy of the sea and the backward state of navigation, exacerbated by the inadequacy and inaccuracy of sea charts.

National jealousies and commercial and political rivalry were, as ever, the underlying causes for many of the century's wars, but war also provided the stimulus for much commercial effort and scientific innovation. Armies equipped with the latest weaponry marched about the countryside, led by generals with only the vaguest of maps at their disposal. At the start of the eighteenth century there were no maps, in the modern sense of the word, anywhere in the world. Indeed, there were plenty of atlases and sketch maps of countries, regions, and districts, but with few exceptions, they were imperfect renditions of nature. Vast gaps in knowledge were filled with speculation and fantastic imagery, as Swift wittily observed:

> So geographers, in Afric maps,
> With savage pictures fill their gaps,
> And o'er unhabitable downs,
> Place elephants for want of towns.[3]

For eighteenth-century sailors aboard the fast-growing fleets of merchantmen and naval ships, the problem of navigating the treacherous seas was compounded by the fact that no one knew how to determine the all-important longitude precisely enough to ensure a safe passage and a welcome landfall. No one knew, with any certainty, the shape of the earth or what lay beneath its surface. Was it hollow or was it solid? Were the Andes the highest mountains on earth or was it the peak of Tenerife? Was the earth a perfect sphere or was it slightly squashed, as Sir Isaac Newton prophesied? Just how did you accurately measure the planet?

The answers to these and a plethora of questions about the nature of the earth, answers we now take for granted, were a complete mystery. Yet, without the answers, maps and sea charts were of dubious value other than at the most provincial level, and provincialism was not in the vocabulary of the politicians, generals, philosophers, scientists, and businessmen of the bustling eighteenth century.

# Contents

# WEIGHING THE WORLD

# 1

# I CANNOT BE WRONG

As the sun rose at the dawn of the sixteenth century, it shone upon a world mostly uncharted, warming newly discovered lands as yet unexplored. Beyond purple horizons, unknown countries and tropical paradises lay slumbering in happy ignorance of the coming storm. In the Old World of the West, the paucity of geographic knowledge had not deterred men from making maps of the World. On the contrary: there was a vast library of maps and atlases, many of which were wildly inaccurate and frequently farcical, showing beautifully engraved continents that did not exist and vague, vast landscapes populated with monsters and cannibalistic savages.

Serene seaways promised wide passages through what were impassable icy wastes that, the cosmographers insisted, led to the riches and spices of the Indies. No one knew from where precisely the spices came, nor did they particularly care. In fact, the strange berries and nuts were grown in the glades of remote East Indian islands and shipped by sea to the coasts of India, from where Moghul traders carried them to Arabia. Arab traders then hauled the baggage overland by camel train through burning deserts to the coasts of the Levant, where Genoese, Italian, French, and English sea traders imported the expensive and shriveled goods into the greedy markets of Europe.

The rich had been satisfied to purchase their spices and exotic goods from the last man in a long chain of traders, that is, until the Ottoman Turks expanded their empire from the east in the fifteenth century, capturing a swath of land stretching from Athens to the Crimea. With Sultan Bayezid II's horde of warriors and warlords controling access to the Danube, Europe's great trade river, and dominating all of eastern Europe, exacting high tolls on goods and traffic, the flow of spices from Asia dwindled. At this juncture, an ancient, much copied map of the world suddenly became very important.

The map was from the *Geographia* of Claudius Ptolemy (fl. 150 A.D.) made at the library of Alexandria during the second century. Much "improved" by Italian cartographers, the map suggested to a young Italian navigator by the name of Christopher Columbus (1451–1506) that there might be a sea route to Cathay and its exotic spices. Columbus reasoned that, the earth being round, he could bypass the Turkish obstruction simply by sailing west until he reached the exotic East.

When Columbus first spied the New World from his flagship, *Santa Maria*, he knew exactly where he was because he had a sea chart. He had discovered, he was certain, the eastern outliers of fabled Japan, gateway to the spice lands. Unfortunately, his chart was hopelessly wrong. The size of the world on which it was based was wrong; his estimate of the distance from Spain to Japan was wrong; the landmasses marked on his chart were wrong. The people he encountered were also wrong—at least, wrong as far as their race was concerned. The natives he met were the cannibal Caribs and their plat du jour, the gentle Arawaks; they were definitely not Japanese or even Chinese. But Columbus did not know this, and there is no reason why he should have. As far as he was concerned, he had been proved right and had found Japan at the very eastern limits of the spice-rich East.

Paolo Toscanelli, a Florentine cosmographer, is supposed to have provided both inspiration and the chart Columbus took with him on his first voyage of discovery. It was based, for the most part, on Ptolemy's ancient map of the world, embellished by the salty tales of coastal traders, fishermen, and an "unknown pilot" who had supposedly seen the fabled lands. Ptolemy's world was the Greek world and was a perfectly round, spherical world. Toscanelli, Columbus, and the natural philosophers of the day accepted this fact almost without question.

From this certain knowledge of a round world, and equipped with the great map, Columbus calculated that his sailing distance west to Japan would be a mere 2,760 miles (4,440 km).[1] In 1492, as his little fleet sailed further and further westward, with no sight of the promised land, Columbus grew increasingly worried, yet he kept his thoughts to himself, confident in his own abilities and having faith in his Florentine map. The crew was frightened and the men were becoming mutinous when, on 12 October (after 36 days at sea), young Juan Rodriguez Ber Mejo saw land from the prow of the *Pinta*.

When Columbus totted up his sailing distance, he realized that they had gone about 4,500 sea miles (8,230 km), considerably further than his original 2,760 miles; the only conclusion the navigator could infer was that the earth appeared to be a lot larger than everyone thought. A few years later, on his third voyage to the Indies (1498–1500), Columbus made an even stranger discovery.

He was observing the latitude by sighting the Pole Star with his quadrant when something very odd occurred. He was certain he knew where he was from his previous voyages, but the latitude observations appeared to be all wrong.

> I found that there between these two straits [the seas between Trinidad and Venezuela], which I have said face each other in a line from north to south, it is twenty six leagues from the one to the other, and I cannot be wrong in this because the calculation was made with a quadrant. In that on the south, which I named la Boca de la Sierpe, I found that at nightfall I had the pole star at nearly five degrees elevation, and in the other on the north, which I named la Boca del Drago, it was at almost seven.[2]

The difference of nearly 2 degrees of latitude for two locations fewer than 70 miles apart could only be explained if the earth, instead of being a perfectly round sphere, had somehow or other manifested some sort of bump near the equator: it was, according to Columbus, deformed.

We might now suggest that the strange anomaly was probably in part the result of his dubious navigational skills and in part to what we would call "atmospheric aberration." But, in 1498, neither Columbus nor any philosopher of the day was aware that the atmosphere behaves like a giant lens, bending light rays. When we watch a big red sun dipping below the horizon on a summer's evening it has, in reality, already set. We see it still because the atmospheric lens bends the light above the horizon. To illustrate the point, if you put a stick in a pond, that part beneath the water appears to be at an odd angle, as the water behaves like a lens, bending the light.

Whatever the cause for Columbus's disconcerting discovery, his thoughts that the earth could be anything other than perfectly round flew in the face of divine perfection; it flaunted the Aristotelian dogma of the church of Rome and challenged the received wisdom of a thousand years. On that starry night in the Caribbean Sea were sown the first heretical seeds of doubt.

# 2

## THE TITAN KING

IN THE 170 YEARS that followed Columbus's discovery of the Americas, a vast library of knowledge about the earth accumulated. Explorers on land and sea, fur and spice traders, and the new merchant companies brought back information on new lands, strange new peoples, and of course, made maps of their discoveries. By the dawn of the eighteenth century, much was known about the shape and size of the earth.

The idea that the world could be a cosmic entity and possess a particular shape and size was a revolutionary Greek concept, and by the fifth century B.C., the world was believed by many philosophers to be round. The Greeks held to the perfection of the mysterious circle with an almost religious zeal; therefore, it was reasoned, the form of the earth had to be a circle that rotated to form a perfect sphere. Although their estimates varied widely and their notion of universal scale was problematic—Anaxagorus of Clazomenae (500–428 B.C.) believed the sun to be very distant, soaring as high above the planet as the Mediterranean was wide—the Greeks were the first to attempt to measure the size of the earth. In the third century B.C., Eratosthenes made the first "scientific" measurement.

Born about the year 276 B.C. in Cyrene, Libya, Eratosthenes, who had been educated at Athens, became librarian of the famous library at Alexandria in 240 B.C. A great scholar and the first "geographer," his wide reading and geographical research led to an interesting discovery. Tradition has it that he learned that near Syene (modern Aswan) was a deep well where the sun at midsummer was reflected in its dark waters. Syene lies very close to the Tropic of Cancer, the northernmost limit of the sun's biannual migration across the equator. Eratosthenes also knew that the distance from Alexandria to Syene, according to official Egyptian "pacers" and from the accounts of camel drivers, was 5,300 stadia (an ancient Greek measure estimated to be between 500 and 607 ft, or 152 and 185 m). Adjusting for the fact that Alexandria was not due north of Syene, he corrected the distance to 5,000 stadia. He then reasoned that if he could measure the angle of the midsummer sun's shadow at Alexandria, by simple trigonometry he could calculate the size of the spherical earth.

One tradition holds that he used an obelisk in the grounds of the library, where he found that the angle cast by the sun at the summer solstice was 7.2°. By dividing this angle into the 360° of a full circle, he calculated that the angular distance from Alexandria to Syene was one-fiftieth of the circumference of the earth. Multiplying the ground distance of 5,000 stadia by 50 gave Eratosthenes a circumference for the earth of 250,000 stadia, or (depending on which definition of the stadion one uses), approximately 24,855 miles, or 40,000 kilometers, a result remarkably close to the modern equatorial circumference of 24,901 miles, or 40,075 kilometers. Modern scholarship[1] suggests that Eratosthenes's work was in actuality far more complex and far more thorough than suggested here; he may even have used an accurate instrument called a *skiotheron,* or shadow catcher, similar to a modern surveyor's transit.

Some years later, another Greek geographer, Posidonius, thought Eratosthenes's circumference was too great and, from his own calculations, deduced a circumference for the earth of about 17,750 miles (28,960 km). It would appear that Posidonius's estimate was the one used by Ptolemy* for his great map of the world, which understated the planet's size by nearly 30 percent. Just why the great cosmographer used such an unreliable figure, when the evidence of his own time pointed to a larger earth, no one is certain. What is certain is that in the 1,400 years after Ptolemy, no one thought it necessary to remeasure physically the size of the planet.

For some 2,200 years, the learned of the West were satisfied that the earth was a perfectly round body. Then, in 1687, the English scientist Isaac Newton postulated that, almost certainly, the earth had to be slightly squashed because of the effects of its daily rotation. Incredibly, in his *Principia,* Newton even calculated the size of the distortion, using the distance between Paris and Amiens, measured by the French scientist Jean Picard.

Among the many physical problems with which the new men of science and philosophy were struggling were the exact dimensions of the earth and whether it was a solid ball or a hollow shell, wherein, some were certain, lay Satan's fiery realm. The more ardent interpreters of Holy Scripture still vigorously defended the process and chronology of the biblical Creation, but slowly and cautiously the accepted dogma of the Catholic Church was probed and tested. There was, of course, no doubt about the age of the world. James Ussher (1581–1656), the

---

*In his *Geographic hyphegesis,* Ptolemy accepts 1° of latitude as the equal of 500 stadia.

bishop of Armagh and a leading expert in his field, accurately calculated the very date of Creation from biblical records. Ussher set the event as having occurred precisely at six o'clock in the evening of 23 October 4004 B.C.[2] There was no evidence or reason to challenge this date as being anything other than a self-evident truth, as many still believe it to be.

Whatever the date for Creation, and despite popular misconceptions, the church never believed the earth to be flat, with Jerusalem at its center. Aristotle, the "virtuous Heathen," had declared it to be round, and round it was. Neither was it any longer believed, although not so generally expressed, that the earth stood at the center of Creation; instead, it traveled around the sun, as stated by the heretic Nicholas Copernicus (1473–1543).

Perhaps we should be thankful that Greek literature and ancient Greek learning was at the very center of Europe's post-Renaissance science. It was fortunate, too, that it had been the ancient Greeks who made the earliest recorded attempts to estimate earth's dimensions. It was even more fortunate that infidel Saracens had the wisdom to preserve what the West, during the Dark Ages, forgot. Had science been based on biblical texts alone, we would have had a narrower pot for science to grow in and might still be dreaming of how we could reach the moon.

With the arrival of the eighteenth century, the thirst for knowledge that had epitomized the previous century exploded. Until Newton's theories about the shape and form of the earth, it was assumed that our planet was perfectly spherical and of a uniform density. A plumb bob suspended above such a perfect world would point directly toward the center of the planet. However, the earth is not perfectly spherical, as Newton postulated, and as we shall see, there is no particular center to where the plumb bob can point.

New theories about the shape and form of the earth abounded and raised many intriguing questions. Newton's laws of gravity set out in his *Principia* were contentious, obscure, and hotly debated, and by no means did everyone agree with the great scientist. If everything, every atom and every particle, created its own gravity, some asked, how large would an object have to be before its effects became apparent? For example, would the mass of a mountain be sufficient to generate its own local gravity? And would it matter anyway?

Some philosophers suggested that an effect caused by mountains, if it existed, might be strong enough to cause a plumb line to be deflected toward the mountain. If so, if the effect were strong enough, it would result in serious errors when measuring the apparent position of the stars. Newton was convinced that the effect did exist but doubted that an instrument could ever be built accurate and sensitive enough to detect the tiny force that became known as "the attraction of mountains." In the event, the effect would turn out to be all too real and lead to some very surprising consequences.

The attraction of mountains, obscure and peculiar as it is, does exist and can be measured, and as we shall see later, it does matter—it matters a great deal. Its greatest effect occurs near the largest mountains, such as the Andes, but even modest-sized mountains can generate enough independent gravity to cause errors. Any instrument that uses the direction of gravity as its reference, such as a surveyor's theodolite or an astronomical observatory's telescopes, will be in error by the amount of the deflection. The most critical measurement affected is latitude.

Even a modest-sized mountain will cause an error of tens of seconds of arc. Now, 1 second of arc in latitude is the equivalent of approximately 100 feet (30 m) on the surface of the earth, so an error of 10 seconds will cause an error of over 1,000 feet (304.8 m). Near giant mountains, the deflection can exceed 1 minute of arc, or 60 seconds, equivalent to more than 6,000 feet (1,829 m) over the ground. Ignoring such errors would lead to inaccurate maps and sea charts. In short, it would lead to the sort of inaccurate maps that led Columbus to discover the New World by chance and many others to become forever lost at sea.

However, at this point in the story, the phenomenon of the attraction of mountains was nothing more than a curious theory and of little consequence. Most maps of the time, that is, all accurate renditions of the earth at a scale sufficient to show useful detail, were surveyed and presented after the Roman model, employing the so-called "Christian topography," and as if the earth were flat. The Romans were practical warmongers and their maps were equally practical. Just because the earth was round was no reason for making things more complicated than necessary.

Mapmaking is an art at least equally as old as writing and probably much older. At Catal Hoyuk, a prehistoric site in Anatolia, a wall painting dated to the seventh millennium B.C. clearly shows a small township and an erupting volcano in recognizable map form. The Babylonians, Egyptians, and Assyrians have left examples of their cartographic skills engraved on clay tablets or picked out on papyrus. Strabo of Alexandria (fl. 20 B.C.) wrote that Anaximander of Miletus compiled the first map of the world in the early part of the sixth century B.C. Even at such early dates, philosophers were struggling to comprehend the shape and size of the world; some followed the Homeric school and believed the world to be disk-shaped; others followed Pythagoras's idea that the earth was spherical.

Much knowledge was lost when Rome fell to the barbarians and religious fanaticism replaced reason and learning in the West. The church descended into a brooding period of dark, cosmic dogmatism that was to develop in a sinister way and persist until the Age of Reason. Fortunately, when western Europe plunged enthusiastically into the Dark Ages, the wisdom of Ptolemy found newer and eager disciples among the thinkers and scholars of Islam.

These Muslim scholars were crucial in transforming the arithmetic of the ancients into modern mathematics. The dissemination, though not the idea, of the rather abstract concept of having a number represent zero is Islamic, as is our "Arabic" numbering system. So enthusiastic and progressive was Islam that places of learning were established. For example, in the ninth century, al'-Mansur, the caliph of Baghdad, founded the House of Wisdom. In its cloisters and gardens, mathematicians like Omar Khayyam developed the processes of calculation called *algebra;* the very word is Arabic. And it was these Arab philosophers who kept alive the learning and perspicacity of the ancients and who added their own new learning and new observations of the cosmos.

Arab scholars also advanced the Greek concept of latitude and longitude, and tenth-century mathematicians such as Abdul Wafa deduced the geometrical principles. The Islamic mathematician and astronomer al'Biruni, a far-sighted individual who believed the earth was not stationary but rotated about its own axis, even made accurate calculations of latitude and longitude for many places in the known world. In 1154, proving that science and learning transcended ideological divides, the Islamic cosmographer al'Idrisi constructed a world map for the Christian crusader King Roger of Sicily.

As the West awoke from its dark slumbers in the fourteenth and fifteenth centuries, it was the Muslim and Moorish Arab scholars who provided the libraries and replenished the lost repository of learning and who passed on their map skills, mathematics, and astronomical knowledge to Renaissance Europe.

However, Europe had not entirely abandoned its mapmaking. In the Middle Ages some magnificent manuscript maps were made in England and France. These works of art were probably produced in some quantity to meet the demands of enthusiastic crusaders who needed to know the way from London to Palestine. Such are the maps of Matthew Paris, a monk of Saint Alban's abbey, drawn about the year 1250, as well as the famous *Mappa Mundi* of about 1300, preserved in Hereford Cathedral. This must have been a most useful map, showing, as it does, the whereabouts of Jerusalem at the center of the world; and almost certainly the beautiful map adorned the hall of a great lord.

Ptolemy's *Geographia* and its world map reappeared in European cosmography circles sometime after 1430. Variations on the map based on Italian models became more widely available with the invention of the printing press in 1450. It was about this time that Europe found that its traditional source of spices and goods from the East via Constantinople (Istanbul) was unexpectedly closed to traffic by the Ottoman Empire. Suddenly, the trade in precious metals, fancy goods, and the all-important spices, so necessary to disguise the flavor of decaying meat, dried up. This event, probably more than any other, stimulated European interest in looking westward as a means of going east and spawned a flurry of exploration.

In what became the age of discovery, the Portuguese set out to explore the coast of Africa, Columbus discovered America, Cabot traversed the bleak shores of Labrador, and Magellan circumnavigated the entire globe. The knowledge these adventurers brought back with them stimulated overheated imaginations and led to the myths of the existence of Northwest and Northeast passages and fabulous shortcuts to the riches of Cathay. What all the explorers agreed on was that the earth seemed to be a lot bigger than previously thought.

Depicting the round earth on a flat piece of parchment was an ancient skill and employed to great effect by cosmographers such as Ptolemy. However, the first scientific method of map rendition came in 1569 when a cosmographer by the name of Gerard de Cramer, or Geradus Mercator,* as he is better known, produced a map of the world whose construction was mathematically sound. Mercator's genius was in employing a device of his own invention (which became known as Mercator's projection) that transposed in a mathematical relationship the roundness of the earth onto the flat plane of a piece of paper.

Mercator (1512–1594) was the eighth child of a poor cobbler in Gangelt, a German border town. While he was still young, his parents died and his elder brother Gisbert, a Catholic priest living in Rupelmonde, Flanders (now Belgium), brought him up. At the age of 16, Gisbert secured Mercator a place as a scholar at the University of Louvain, where he studied the teachings of Aristotle. It was at Louvain that his sympathies for Protestantism first became apparent and where he began to feel that the views of Aristotle, whose teaching was enshrined in Catholic dogma, were misconstrued. However, such thoughts in the early years of the Reformation were distinctly and dangerously heretical, and he kept them to himself.

After a short period of European travel, in 1534 Mercator returned to Louvain to study mathematics and cosmography under the celebrated Dutch astronomer and mathematician Gemma Frisius (1508–1555). In 1537, he produced his first major map, of Palestine, and 3 years later he made a map of Flanders that he constructed from a survey he personally supervised.

Then, in 1544, in a wave of arrests, Mercator and many of his friends and acquaintances were charged with heresy. After 7 months' incarceration in the dis-

*Cramer = Dutch for "merchant," which, Latinized = Mercator.

Geraldus Mercator. Courtesy Library of Congress.

mal cells of Louvain castle, he was released without charge; many of his associates were less fortunate. The women were buried alive; the men were burned at the stake or beheaded.[3] It was a frightening lesson: Mercator learned that these were dangerous times for mapmakers who flirted with concepts that could so easily be interpreted as devilish. Impoverished and desperate, Mercator and his family fled to Duisburg in the Duchy of Cleves, a progressive Calvinist town where attitudes were a little more liberal.

In 1552, Mercator set himself up in business, making globes, atlases, and "large-scale" maps of Europe, including the British Isles. It was in his shop at Duisburg that he perfected his mathematical projection for the earth and where he produced his famous world map. Duisberg was also where he coined the first known use of the word "atlas" to describe very small-scale maps of countries, "to honour," he wrote, "the Titan, Atlas, King of Mauritania, a learned philosopher, mathematician, and astronomer."[4]

Mercator's mathematical projection of the world revolutionized mapping and for the first time allowed the accurate construction of maps and charts. Fortunately, it is not necessary to understand the mathematics to see how his projection works. It is easily visualized as a piece of photographic paper wrapped as a cylinder around the equator of a glass globe, one that has engraved on it the lines of latitude and longitude. If a light is placed at the center of the globe, the lines of latitude and longitude are "projected" onto the paper. When the paper is unwrapped and developed, the curved lines of latitude and longitude appear

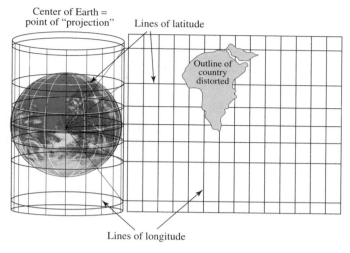

Center of Earth = point of "projection"

Lines of latitude

Outline of country distorted

Lines of longitude

Mercator's projection of the earth.

on the flat paper as straight lines. The spacing between the latitude lines increases toward the poles, which is a drawback to this sort of device, and Mercator's projection runs out of steam in the high latitudes. But this would not be a problem until explorers ventured to the icy polar wastes many years in the future.

The latter part of the sixteenth century witnessed an extraordinary increase in the number of maps and atlases in production, spurred by the new discoveries and the demands of sea traders and merchants. The worldly imaginings of Ptolemy finally faded away in 1570, when Abraham Ortelius, geographer to Phillip II of Spain and a close friend of Mercator, published *Theatrum Orbis Terrarum,* an atlas of the world composed (almost) entirely from contemporary explorers' knowledge and observations.

Fifteen years later, Mercator's mighty map of the world, his *Atlas in Three Parts,* was published in Amsterdam. It was, at the time, the most accurate rendition of the earth ever made. Many errors of earlier times were corrected and the latest discoveries included. For the first time, the width of the Mediterranean Sea was shown close to its modern value, 10 degrees of latitude less than on Ptolemy's map.

Between 1572 and 1618 Braun and Hogenberg, in the Rhineland, produced plans of over 500 towns. In England, Christopher Saxton began his survey of England and Wales, publishing his *Atlas* in 1579, a work unsurpassed in accuracy and quality for nearly 200 years. The cartographer John Speed published his *Atlas* in 1611, which contained superlatively detailed town plans, the first of their kind, and the boundaries of the ancient administrative sections called "hundreds" and "lathes." With various amendments and additions, Speed's maps remained popular until the close of the eighteenth century.

In 1675, the cartographer John Ogilby published *Britannia,* the first atlas of Britain to show the main postal roads. Yet only the atlases that depicted large areas of a country or a continent employed Mercator's projection as a means of preserving accuracy by rendering the curved surface of the earth on to paper. All other maps were drawn as if the world was flat. But, for accurate mapmaking of large areas at anything greater than atlas scales, removing or allowing for the distortions caused by the roundness of the earth was a prerequisite.

Mercator's projection made it easy for maps and charts to be drawn very accurately, but the instruments to measure the all-important latitudes and longitudes in the first place had not similarly advanced. Navigating at sea and determining an accurate geographical position on land depended entirely, as it had for thousands of years, on observing the stars. Apart from the Pole Star, very few star positions were accurately known for anything but the most basic sort of navigation. It was a sort of Catch 22: latitudes depended on knowing the positions of the stars, but knowing the position of the stars depended on making accurate observations of their locations from the earth in the first place.

To overcome this difficulty and address the paucity of data, well-equipped and properly staffed astronomical observatories were needed to measure precisely the places of the stars and to follow the motions of the planets and the sun and moon. The first modern telescopic observatory charged with charting accurately the night sky was built in Paris in 1667. Britain followed suit in 1675 with its Greenwich Royal Observatory.

Of the two critical geographical measurements, latitude was relatively easy to determine by using a primitive quadrant or astrolabe to observe the "altitude" of a star. Longitude, on the other hand, was totally abstract; it required an origin, a physical place, from where it could be counted. The French naturally counted from Paris, the British from Greenwich, and the Spanish from the Canary Islands. Unlike the astronomical nature of latitude, finding the longitude required a "mechanical" solution. Stimulated by a large cash prize, the problem was eventually solved by the British clockmaker John Harrison with his remarkable invention of the chronometer. However, prizes for solving the longitude problem had a long history that well demonstrates the commercial and military importance that merchants, politicians, and scientists placed upon it.

The maritime nations of Spain and Portugal were among the first to recognize the importance of longitude for the safety of their shipping and, in particular, mapmaking and exploration. Toward the end of the fifteenth century, these two neighbors were in dispute over their overlapping exploration and commercial interests. In 1493, the year Columbus brought back news of his discovery of "Japan," Pope Alexander VI (a Spaniard) issued a Bull of Demarcation to settle the argument.

Alexander, a notable pope but less so a cartographer, drew a line of longitude on a sea chart 100 leagues (400 Roman miles) west of the Azores. Everything undiscovered to the west (including the New World) he bequeathed to Spain, and everything to the east went to Portugal. In the pope's mind, and in the minds of his advisors, the solution was simple and expedient. What they failed to realize was that it was almost impossible to implement the bull with even the remotest degree of accuracy and that their line of longitude would cross the poles and run down the other side of the world.

In 1567, King Philip II of Spain announced a cash prize for the first person to solve the longitude problem; in 1598, his successor, Philip III, raised the stakes further. Nearly 60 years earlier, in 1530, the Dutch mathematician Gemma Frisius, who had taught Mercator, had put forward his method of finding the longitude by using a clock. His suggestion was that a clock would be set ticking at the point of departure, and "while we are on our journey," wrote Frisius, "we should see to it that our clock never stops. When we have completed a journey of 15 or 20 miles, it may please us to learn the difference of longitude between where we have reached and our place of departure. We must wait until the hand of our clock exactly touches the point of an hour and at the same moment by means of an astrolabe . . . we must find out the time of the place we now find ourselves."[5]

Observations with an astrolabe, an ancient device for measuring the altitude of a heavenly body above the horizon, were commonly used to determine local apparent time; in theory, his proposal could have worked but for two facts. First, astrolabes were not particularly accurate, and second, seeing "that our clock never stops" was hopelessly optimistic. There was not a clock in the whole world that was either accurate enough or reliable enough. Despite the absence of the technology, Frisius's method did, in the end, triumph but only after 230 years.

The Spanish longitude prize was worth an incredible 6,000 gold ducats plus a pension of 2,000 ducats for life to the discoverer. One early respondent was the Italian astronomical genius Galileo, who wrote to the Spanish court in 1616 with a clever plan. Instead of relying on the fallibility of man-made mechanics, Galileo proposed using the regular and predictable heavenly mechanics of the moons of Jupiter, which he had discovered in 1610 and for whose motion he had compiled tables. In essence, Jupiter's tiny moons replaced the hands of Frisius's clock; in all other respects the solution was the same.

For 16 years, Galileo tried in vain to persuade the Spaniards of the excellence of his scheme. Then, in 1636, the States General of the United Provinces of the Netherlands joined the longitude race and offered a prize of its own. Galileo, ever on the lookout for cash, quickly turned his attention from the reticent Spanish to the enterprising Dutch. The States General set up a commission to investigate his proposal, but by then the great Italian was under house arrest by the Inquisition for his heretical belief in Copernicus's theory of a sun-centered universe.

# 3

# A CALM AND GENTLE CHARACTER

ONE OF GALILEO'S VEHEMENT DETRACTORS was the astrologer Jean-Baptiste Morin (1583–1656), a wan-looking man with a shock of wild hair; his broad forehead ranged above a pair of troubled eyes. Morin passionately believed the earth to be fixed firmly in space, as befitted the church's Aristotelian ideal, and he renounced any philosopher who challenged him.

Morin was a delightfully arrogant and self-opinionated person: "I am excessively inclined to consider myself superior to others on account of my intellectual endowments and scientific attainment,"[1] he wrote of himself. He had been born into a wealthy family of Villefranche and educated in philosophy and medicine. At the age of 30, he was sent on an investigative trip to Germany and Transylvania to study the mines and minerals of the area for the bishop of Boulogne. This was an interesting journey and stimulated within Morin a curiosity for the mechanical sciences and an overdeveloped opinion of his own righteousness. In 1630, he became a professor of mathematics at the College Royale in Paris, where he embarked on a distinguished astrological career.[2]

In the seventeenth century, the gap separating astrology from astronomy was narrow to the point of nonexistence; it is no surprise, then, that Morin contributed so much to the junior science while being a master of the senior. In 1634 he approached Cardinal Richelieu (1585–1642), King Louis XIII's chief minister, with a method for finding the elusive longitude. His proposals were an advance on the earlier work of the Nuremberg cosmographer Johann Werner (1468–1522), whose idea in 1514 was to observe "the distance between the Moon and one of the fixed stars which diverges little or nothing from the ecliptic."[3] King Louis's advisors set up a royal commission of eminent scientists, including the celebrated mathematicians Etienne Pascal and Clause Mydorge, to study Morin's proposals.

Like Werner, Morin's idea for finding the longitude was to take advantage of the moon's passage across the celestial sphere. Morin claimed that by observing the angular distance of the moon from the sun or from certain charted stars, he could use this information to derive a longitude difference. Where he differed most from old Werner was that Morin included corrections for atmospheric refraction and for the "lunar parallax," the difference in angle made by the moon when viewed from different points on the earth. Werner's proposal, known widely as the "lunar distance method" would, in the distant future of Morin's time, rank as important a solution to the problem as John Harrison's chronometer. Morin had little faith in clocks, recalling the mechanical ideas of Frisius: "I do not know if the Devil will succeed in making a longitude timekeeper, it would be folly for man to try."[4]

Instead, Morin proposed taking advantage of the recent improvements in optical instruments to make his lunar method a practical one for simple seafarers. He even suggested erecting an observatory to determine accurately the necessary lunar data. For 5 years, the committee met on and off to discuss and evaluate Morin's proposal. For 5 years Morin argued with the committee; he even argued with the great philosopher René Descartes. But the mathematics of the solution were far too complex for seafarers to solve, and the data on the moon's erratic movements, vital for the theory to work, were too difficult to measure or predict.

Eventually the scientific community in France could stand Morin's bickering and argumentative ways no longer and sent him away to spend his declining years in isolation, a scientific pariah, absorbed in his astrological studies. His original benefactor, Cardinal Richelieu, died in 1642, but his successor, Cardinal Mazarin, ultimately awarded the old astrologer his 2,000 livres longitude prize, even though his ideas had been set aside.

The lunar distance method was shelved, and no further progress on the longitude problem was made in France until 1661, when King Louis XIV appointed the far-sighted Jean-Baptiste Colbert his minister for home affairs. Colbert recognized that what was needed, not just for the longitude but also for a range of scientific advances, was a college or foundation made up of all the greatest minds in Europe. In 1666, he founded the Académie Royale des Sciences and persuaded the king to fund it from the royal purse. The society had a wide remit to study all things scientific and to provide philosophical advice to the government. Its most important directive was, however, to give France better maps and sea charts and to improve the dangerous business of navigation.

Colbert was totally committed to his new society and offered large remuneration to any scientist willing to dedicate himself to its work. The greatest brains of the day were invited to join, including Gottfried Leibniz, Ehrenfried Tschirnhaus, and Isaac Newton. The Dutch physicist, astronomer, philosopher, and mathematician Christiaans Huygens, inventor of the pendulum clock in 1656 (or 1658), and the astronomer Romer decided to take up the offer. The astronomer and mathematician Abbé Jean Picard, the optical instrument maker Adrian Auzout, and many other scientists soon joined them. The astrologers set the auspicious winter solstice of 22 December 1666 as the date when the Académie Royale opened for business.

Harking back to the days of Frisius, Huygens was convinced that the longitude problem could be solved by the development of a clock of his own invention. He set himself the task of producing a marine timekeeper accurate and reliable enough to keep regular time for months on end, in any climate, and regardless of a ship's violent motion. For 20 years, Huygens kept trying and experimenting but to no avail. The mechanical solution would have to wait another 100 years. Huygens is best remembered today for his work on rotating bodies and for his theories on light, theories that remain some of the most brilliantly original contributions to modern science.

The society's astronomers were provided with a house near Cordeliers and another at the Louvre to set up and operate their instruments. A fine quadrant was acquired together with a large sextant and a precisely engraved Moorish sundial. Astronomical observations began in January 1667. It soon became clear that the observing sites suffered from practical faults, and Colbert persuaded the king to fund France's first official observatory. Located at Faubourg on the south bank of the Seine, the famous Observatoire de Paris began its groundbreaking work on the day of the summer solstice, 21 June 1667, although the buildings were not completed until several years later.

Samuel Johnson visited the observatory in 1775, finding "a large building of a great height. The upper stones of the parapet very large, but not cramped with iron. The flat on the top is very extensive; but on the insulated part there is no parapet. Though it was broad enough, I did not care to go upon it. Maps were printing in one of the rooms."[5] The meridian (north-south line) that the astronomers determined passed through their observatory became the official meridian line of France and for many years was the origin for all French longitude measurements.

So important was a solution to the longitude problem, and so desperate the Académie Royale to discover it, that it invited Jacques Graindorge, a Benedictine monk from the abbey of Fontenay, to reveal his divinely inspired solution. In a letter to Colbert, Graindorge had claimed that his solution was as simple as finding latitude (which was, indeed, quite simple). In January 1669, Graindorge attended a secret meeting with the academicians and revealed his secret. Huygens and Jean Picard were asked to evaluate the idea and quickly pronounced it

The Paris Observatory at the time of Cassini.

ridiculous; what it was we do not know. Despite their conclusions, the académie decided wisely to acknowledge the power of divine revelation and granted the monk 1,200 livres for his trouble.

By this time, the Académie Royale had returned to Morin's lunar distance method for the longitude. In principle, the solution was elegant and practicable. The problem was that the motion of the moon in its 18-year cycle is extremely complex and, at the time, defied prediction. Beginning in 1667, the Paris astronomers spent 2 years making countless observations of the moon's position among the stars in an attempt to find the secret to the riddle. Unable to solve the difficult business of lunar forecasting, the academicians' studies reinforced earlier conclusions that the mathematics were too abstract and would, in any case, prove impossible for mariners to use in practice. However, Jean Picard, the académie's leading astronomical mathematician thought he saw a possible solution.

Jean Picard (1620–1682) was a quiet and modest Jesuit priest, a prior from Rillé in Anjou. Known as "the father of French astronomy," Picard's introduction to astronomy came in 1645, when he observed a solar eclipse. In 1655 he was appointed professor of astronomy at the College de France in Paris, from where the fame of his considerable skills and knowledge led to his becoming one of the earliest recruits to Colbert's académie. With regard to the longitude, Picard had been in correspondence with an Italian mathematician by the name of Jean Dominique Cassini, who, it seemed, had an alternative solution to the problem. From his observatory at Bologna, Cassini had steadily built on the pioneering work of Galileo, observing the eclipsing moons of Jupiter and compiling precise tables for predicting their regular motions, which he published in 1668.

Picard held Cassini in the highest regard, although he had never actually met him, and persuaded Colbert to ask King Louis to invite the Italian astronomer to become the director of the new Paris Observatory. In spring 1669, just a few months after receiving his letter from the king and with the blessing of his benefactor, Pope Clement IX, Cassini arrived in Paris with his bags, trunks, and astronomical instruments. As things turned out, Picard came to rue the day he recommended Cassini because the Italian proved to be something of an envious and truculent colleague who went to great lengths to thwart Picard's ideas for improving the observatory.

Jean Dominique Cassini (1625–1712; Italian: Gian Domenico Cassini) was born at Perinaldo in Italy and studied at the Jesuit College of Genoa. Like Morin, his initial studies of the stars were for the pseudoscience of astrology, but this quickly developed into a devotion to the practical science of astronomy. In 1644, at the age of 19, Cassini left home to study at the University of Bologna, where the wealthy Marquis Malvasia was building an observatory. Six years later, his reputation growing, he was appointed to the university's chair of astronomy. A mathematician, as well as astronomer, and a man of extraordinary intellect, Cassini also studied geography and hydraulic engineering and was named by Pope Alexander VII as his inspector of water and waterways.

Despite Cassini's tables for accurately predicting the motions of Jupiter's moons, which allowed very accurate time measurements to be made, the riddle of longitude still needed another vital piece of information. The size and form of the earth were still uncertain: only one scientific experiment had been conducted to measure its circumference, by Willebrord van Roijen Snell (1580–1626), in the early years of the seventeenth century.

Snellius, as he is better known, was the son of a wealthy professor of mathematics at the University of Leiden in Holland, where young Snell himself studied law. But it was mathematics that intrigued him, and through his studies, he made considerable contributions to science. He is probably best remembered today for the optical law of incidence and refraction named for him. However, the most important of his works published in his lifetime was *Eratosthenes Batavus,* and it revolutionized the science of earth measuring. In 1615, emulating the principles of Eratosthenes 1,900 years before, Snellius had measured the length of a meridian arc by using a chain of thirty-three triangles, observed from the tops of churches, across the fields and hedgerows of the flat Dutch landscape. He had then measured the latitude at the north and south end of his

Jean Dominique Cassini
(1625–1712). Courtesy Library
of Congress.

chain of triangles. By calculation, assuming the earth perfectly spherical (as did all philosophers in those days), Snellius worked out a diameter for the earth that was just 3.4 percent smaller than modern values.

Although not the first to use triangulation (that claim rightly belongs to Gemma Frisius 60 years earlier), Snell explicated the idea of measuring long distances over the ground by measuring the angles in a chain of triangles rather than by measuring the distance with a surveyor's chain. The method was also far more accurate and significantly speedier than physically measuring directly across the land. Snell's experiments, funded and aided by his patron, Baron Sterrenberg, demonstrated the accuracy of the method and proved conclusively that a framework of stations so developed could be used to control a mapping project.

Fifty years later, in 1669, the mild-mannered and brilliant Abbé Picard began the great task of accurately measuring the size of the earth for the Académie Royale. Resorting to Snell's triangulation methods and the principles of Eratosthenes, Picard established a chain of thirteen triangles covering the meridian between the "Pavilion" at Malvoisine, south of Paris, and the church tower at Sourdon and onward to Amiens. For measuring the horizontal angles in his tri-

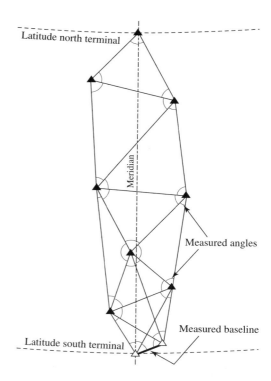

Latitude north terminal

Meridian

Measured angles

Measured baseline

Latitude south terminal

Principles of triangulation and
the meridian arc.

angulation scheme, Picard employed a *quart de cercle,* a horizontally mounted quadrant with a radius of more than 3 feet (90 cm). The *secteur astronomique* he used for measuring the latitude at the terminal points also had a radius of 36 inches (90 cm) and was probably the first astronomical instrument to be fitted with a telescope. Picard is also credited with being the first person to use very fine cross-wires set at the focus of a telescope to aid in precise measurements of star positions. The scale arc of the instrument was so well graduated that the observer could read angles to one-quarter of a minute of arc (1/240 degree).

Two baselines were set out and measured: one of 5,663 *toise du Châtelet* (an ancient French measure also known as the French fathom, about 6.4 ft, 1.95 m) between Villejuif and Juvisy, southwest of Paris, and a verification base near Montdidier, southeast of Souron. The triangulation scheme, with its baselines, provided Picard with the linear distance between the two terminals. Telescopic observations of Jupiter's eclipsing moons provided the absolute times. Two of Huygens's new-fangled pendulum clocks were used to measure relative time so accurately that the clocks, wrote Picard, "marked the seconds with greater accuracy than most clocks mark the half hour."[6] After long study and detailed calculations, Picard was able to announce that the length of a degree of latitude

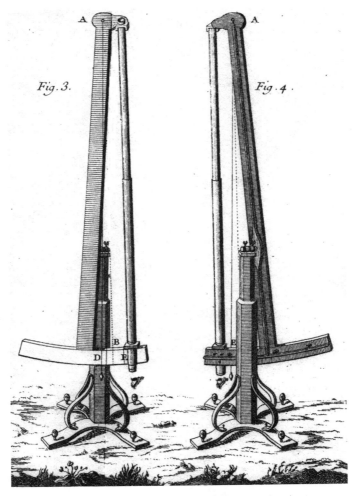

Zenith sector used by Picard. From *Mesure de Dégré du Meridien en France,* Mémoires de l'Academie Royale des Sciences, Paris, 1736.

north of Paris was 57,060 toises (111.3 km). From this measurement, Picard was able to calculate that the diameter of the spherical earth was 7,925 miles (12,750 km). This was the value sent to Isaac Newton, which the great scientist used to formulate his laws.

At the time Picard made this, the first measurement for the size of the earth using scientific principles and precise instruments, no perfect "standards" of the toise had been preserved; many had been lost or mislaid. Picard recognized the importance of conserving a correct standard of length and lighted on the ingenious idea of preserving the toise in a naturally replicable way, a precursor to the way measures would be preserved in the distant future of our own time. Picard knew that the pendulums of Huygens's clocks had to be exactly the right length to beat the seconds precisely and constantly. Any variation by the smallest amount would cause the clocks to lose or gain time. Therefore, Picard reasoned, to preserve the length of his toise all he had to do was compare its length to the pendulum of the observatory clock that was beating the seconds so that it could be reproduced exactly at any future date.

When Cassini was appointed director of the Paris Observatory in 1669, one of the first tasks he set himself was to compile the volumes of cartographic information the académie had acquired and to augment this with all the new information then being gathered from around the world. This information he then corrected, using the académie's new knowledge of the size of the earth, to produce the first accurate, "scientific" map of the globe on which was plotted only those geographical locations the latitude and longitude of which had been properly observed. The great map of the world, 24 feet in diameter (7.3 m), which Cassini called his "planisphere," was laid out on the floor of the Paris observatory with the heretical North Pole displacing Jerusalem as the center of Creation.[7] The map became a Parisian wonder; it was visited by the king and attracted the attention of scientists and nobles throughout France.

In 1673, 4 years after his arrival in France, his fame now considerable among the men of philosophy, the king of France conferred French nationality on Cassini to ensure that the Italian remained permanently in France. Shortly after, Cassini cemented his new nationality by marrying a French woman, Geneviève Delaitre. Despite his dictatorial and dogmatic personality, unfortunate traits that led to serious differences with Picard and others of his colleagues, Cassini was by nature a "calm and gentle character" and deeply religious, as befitted his Jesuit upbringing.

Cassini was first and foremost an observer, and his contributions to science were great and many, including determining the orbital periods of Jupiter, Venus, and Mars; discovering four new Saturnine satellites; and developing a theory for the motion of comets. Like Newton, Cassini thought the world was not a perfect sphere but was, to some degree, slightly distorted. Picard was of a similar persuasion, but where he differed strongly with Cassini was that he, like Newton, believed the earth to be flattened toward the poles.

To settle the dispute over the true shape of the earth and advance its understanding of the world, the Académie Royale des Sciences in 1672 dispatched the astronomer-mathematician Jean Richer and the 52-year-old Jean Picard to Cayenne, in French Guyana, to conduct a series of experiments. The main objec-

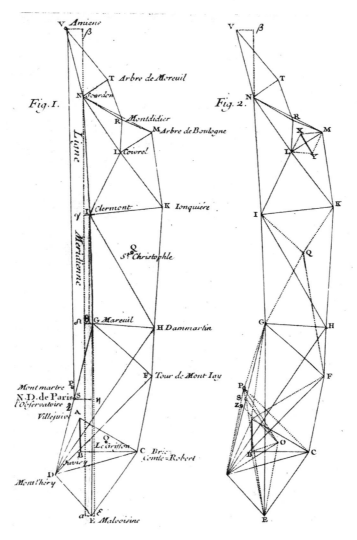

Picard's meridian arc. From *Mesure de Dégré du Meridien en France,*
Mémoires de l'Academie Royale des Sciences, Paris, 1736.

tive of the expedition was to observe an opposition of Mars, our nearest planetary
neighbor, and to determine the local longitude and strength of gravity. Picard
took with him one of Huygens's pendulum clocks, carefully calibrated against
the astronomical clocks that clanked away the seconds in the Paris Observatory.

The theory behind the gravity experiment was simple but clever: the strength
of gravity is greater the closer one is to the center of the earth; hence, if the earth

were anything but perfectly round, the measured strength of gravity should differ from place to place. As it happened, the clock so carefully calibrated in Paris, instead of remaining steady or speeding up, as Cassini expected, began instead to lose time at the rate of 2.5 minutes a day, and the French scientists had to shorten its pendulum by 2.8 lignes to get it to run correctly.[8] The conclusion Picard drew from their experiments was that Cayenne was *further* from the center of the earth than was Paris, and therefore the earth was oblate, flattened toward the poles, as Newton had postulated. In Paris, Monsieur Cassini was not convinced and laid the blame for the "error" on the quality of the observations and the laxity of the observers, which conclusion only went to further exacerbate his degenerating relationship with the aging Picard.

Another expedition was sent out in 1681, this time to the island of Gorée off Senegal in West Africa, a slaving post recently taken from the Dutch. Two eminent astronomers, Charles-Nicolas Varin and Louis des Hayes, were chosen to lead the expedition and were trained personally by Cassini on how to make the observations to ensure that they were not corrupted by the "flawed" methods of Picard and Richer. To derive the longitude, Galileo's method of timing the eclipses of Jupiter's moons, which Cassini had perfected, was used, and the expedition was equipped with two long-case pendulum clocks to keep precise time. One clock was calibrated to keep mean time and the other to maintain sidereal, or star, time (being 3 minutes, 56 seconds shorter per day than mean time). Like Richer and Picard before, Varin and des Hayes discovered that, despite Cassini's careful tutoring, their clocks refused to run as when calibrated in Paris, and they, too, had to shorten the pendulums.

When the intransigent Cassini learned of the problem, instead of accepting that the observers were correct and his own thoughts suspect, he blamed the clocks' mechanics. However, when the news of the experiments reached Isaac Newton in England, the knight was delighted for he knew that it was the diminished strength of gravity near the bulging equator that caused the clocks to run slower. Sadly, Jean Picard was no longer around to learn of the news, having died on 12 October 1682.

In 1683, Cassini himself began working on an extension of Picard's meridian arc. Assisted by his son Jacques and a Monsieur La Hire, he extended the triangles north to the church tower of Dunkerque, on the northern coast of France, and south to Perpignan, nestling at the foot of the high Pyrenees. He employed exactly the same methods as Picard, using astronomical instruments of large radius to get precision and the eclipses of Jupiter's moons for absolute time. When the observations were computed, Cassini discovered to his great satisfaction and the consternation of others, that 1 degree of latitude on the north side of Paris was the equivalent of 56,960 toises (111.1 km) and south of the capital, 57,097 toises (111.3 km). Instead of getting longer as he went north, the degrees were getting shorter, just as he had predicted they would. The only conclusion that he could draw from this remarkable difference was that the Picard-Newton "oblate" spheroidal earth hypothesis was wrong and the earth was, as he suspected, a "prolate" spheroid with a polar axis slightly longer than the equatorial dimension, a bit like a lemon in more ways than one.

From his interpretation of the results, and despite the evidence from the equatorial expeditions and Newton's theoretical calculations or Picard's observations, Cassini could not accept that his meridian arc might contain a small error, arguing, with some justification, that his toise was slightly different from that used by Picard. The difference between the north and south segments of his arc represented 650 feet (198 m), the equivalent of just 8 arc seconds of latitude, and in the closing years of the seventeenth century there was no instrument that could make such a delicate measure. It has been suggested that the most likely explanation for the strange result was that, instead of observing the moons of Jupiter, Cassini had mistakenly observed those of Saturn. This is entirely plausible because, throughout 1683, Saturn and Jupiter were moving together, often in conjunction, between the constellations of Leo and Cancer. The quality of the low magnification optics of his instruments, which suffered from the distorting rainbow effects of chromatic aberration, was probably insufficient to discriminate the faint distinguishing rings of Saturn when their edges faced the earth.

Cassini's misconception, whatever the cause, was to persist for many years, in part because of the great reputation he enjoyed in scientific circles and in part because of the work of his son, Jacques Cassini (1677–1756), who succeeded to the directorship of the Paris Observatory on his father's death in 1712. Both father and son, brilliant as they were, were traditionalists. They could neither accept Kepler's laws of planetary motion, which displaced the classical harmony of the ancients, nor bring themselves to believe in Newton's gravity theories, let alone his arguments for the shape of the earth. It was not until the reign of the third of the Cassinis (César-François Cassini de Thury, 1714–1784) that the great family was finally able to come to terms with the truth of the matter.

# 4

## THE GALILEO OF FRANCE

AFTER SOME 2000 YEARS of classical perfection, it was at last widely accepted that the earth was a less than perfect, slightly distorted sphere. The Cassinis remained convinced from their extensive measurements that the distortion was "prolate," with a polar axis a few miles longer than the equatorial axis. Others were just as certain that the Newton-Picard-Huygens hypothesis for an oblate spheroidal earth, supported by Picard's somewhat tenuous observations and the results of the overseas gravity experiments, was the correct one. It was a classic example of theory being more reasonable and more attractive than actual scientific fact—a view that a modern scientist might find difficult but a politician admirable.

In the early years of the eighteenth century, the overland measurements to detect the dimensional distortion of the earth had been performed only on the soil of France, and the results had led to contention and bitter divisions rather than to concord and harmony. If the Newtonian school were to be vindicated, Picard's experiments would have to be repeated at other, more extreme localities. Early ideas for determining the dimensions by making measurements of longitude were still popular and frequently debated, that is, until 1733, when Giovanni Poleni (1683–1761), professor of mathematics and astronomy at the University of Padua, demonstrated in a prize-winning paper that measurements done on the meridian would be far more reliable. Polini's ideas were promoted by the Newtonian champion Pierre Louis Moreau de Maupertuis (1698–1759), "the Galileo of France," as Voltaire called him.[1]

To resolve the issue, the engineer Louis Godin des Odanais proposed a solution to the Académie Royale: he would lead an expedition to Peru to measure the length of a degree in the meridian at the equator. After much debate, the académie voted its agreement. To make the most of the opportunity, the expedition would not be confined to the measurement of the earth alone but would be a wide-ranging and ambitious scientific undertaking. Funds were secured, and resolving the complexities of the logistics and selecting and agreeing who would be included in the expedition continued until early 1735. On 16 May, amid great celebration, Godin's expedition set sail for Peru. However, it was not to be the first on the ground.

Ten days after Godin's departure, Maupertuis presented a memoir advocating his own "northern" expedition to determine if a meridian arc toward the North Pole would be longer than the Picard-Cassini arc of France. Given the great distance Godin had to travel to Peru, Maupertuis thought he would have an answer from the North Pole a year or two sooner. After much argument and dissent, the council of the Académie Royale agreed to fund the extra expedition to see if the same effect that was discovered across France would be found at the earth's extremes.

Maupertuis's idea was to conduct his experiment along the island archipelago off the northern coast of Norway. Fortunately, one of Europe's leading astronomers, Anders Celsius (1701–1744), was in Paris at the time. Celsius, best remembered for the thermometer scale he devised, was professor of astronomy at the University of Uppsala in Sweden; he had been following the discussions on the shape and form of the earth and was a leading advocate for the polar meridian arc measurement.[2]

Instead of Norway, Celsius persuaded the academicians to select the Grand Duchy of Finland, his homeland's northern vassal and buffer state with Russia. The undulating terrain of Lapland, in the far north of the duchy; the existence of a simple but efficient rural economy; and the nature of the Finnish political situation, coupled with the eloquence of Celsius's arguments, were sufficient to persuade the French philosophers.

Maupertuis was to lead the expedition. A mathematician, astronomer, and mechanicist elected to the académie in 1731, he was a genial but obdurate man. Every inch the epitome of the French savante, Maupertuis was agnostically inclined and a firm disciple of Isaac Newton, holding fast to the great English scientist's belief in a solid and oblate earth. It was Maupertuis's strong character and unshakable faith in Newton's gravity theories that persuaded many in the académie to accept these radical ideas. The expedition to Lapland was to include Celsius, who would act as his country's official observer, guide, and interpreter. Also politically significant, the inclusion of Celsius in the expedition was a sign of France's recognition of Swedish academic excellence and, more crucially, its status as an ally.

On 2 May 1736, the expedition to test Newton's theory for the shape of the earth set out by sea from Dunkerque, bound for Finnish Lapland. Maupertuis's team comprised Charles Étienne Louis Camas; Pierre-Charles de Monnier; Alexis

Claude Clairaut; the académie's correspondent, l'abbé Réginal Outhier; a secretary by the name of Sommereux; Herbelot, the expedition's draftsman; and, of course, Anders Celsius. The French scientists arrived at Stockholm, the capital of Sweden, in early June and presented their credentials to the authorities. Here, Anders Hellant was conscripted into the team as general factotum and the expedition's translator of the difficult Suomi (Samë) tongue of the northern Laps. After a few days rest, the nine-man team set off across the Gulf of Bothnia on the last stage of its 1,500-mile journey. Their objective was the town of Torneå at the mouth of the Torneonjoki (Tornio River), that today marks the border between modern Sweden and Finland.

Arriving on the day of the summer solstice, 20 June 1736, and with no time to lose, the scientists began immediately, spending 16 days in conducting a thorough reconnaissance of the area and identifying the sites for the survey stations. Once again, Snellius's pioneering methods were to be employed to carry a chain of triangles from Torneå northward across the wild landscape of Lapland. However, Torneå lies fewer than 50 miles south of the Arctic Circle, and if the French scientists were to measure a full degree of latitude, they would end up some 20 miles north of it. In the high latitudes of the land of the midnight sun, summer is 73 days of continuous light. On the other hand, winter comes suddenly and brings sunless days on end. If the expedition were to achieve its objective, the observation of the great triangles, which needed full daylight, had to begin immediately. The all-important baseline measure could wait until later.

The Torneonjoki flows more or less north to south in a series of long, narrow lakes joined by swift-moving streams full of white-water rapids and other dangers. On either side of the river's wide valley are low, rounded hills of ancient glacier-smoothed stone, covered for the most part by lonely, wolf-haunted forests of pine and spruce. Maupertuis's plan was to use the high points offered by these hills to carry his triangles northward toward the peak of Kittisvåra, some 70 miles (134 km) north of Torneå. To measure the angle at the apex of each triangle, the expedition was equipped with a 2-foot radius quadrant mounted horizontally on a cast-iron stand. It was a very fine but heavy instrument, one of the most accurate ever constructed, and the errors in the triangles it measured never exceeded 12 seconds of arc (1/300 of a degree).

The Swedish government provided Maupertuis with a contingent of soldiers under the command of Colonel Rietz to help with the more mundane aspects of the work, such as erecting beacons and hauling heavy equipment, and to act as guides and protectors. With the assistance of their military helpers, the French scientists began to set up the hilltop beacons, completing the work on 18 July.

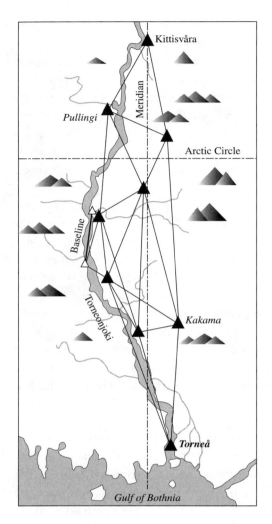

Maupertuis's meridian arc in Lapland showing the triangles and terminal points.

The next day they set up their large quadrant at the pretty wooden church of Suensåri in the center of Torneå and sighted across the dark green forests to the beacons on the summits of Nivavåra (408 ft, 124.4 m) and Kåkamavåra (619 ft, 188.6 m), 20 miles distant.

Soon the temperature on cloudless days soared to 80°F (27°C). The air in the dense forests was heavy with the scent of pine and infested with clouds of whining, biting mosquitoes that made the delicate observations a painful, tedious business of swollen eyes and itching flesh. Summer passed into fall, which brought with it the first snowy harbingers of the fierce northern winter. It was 2 September by the time the astronomical party reached the peak of Kittisvåra

(524 ft, 156.7 m) above the little village of Pello, marking the end of the first phase of the work. Including the starting point at Torneå, Maupertuis had required just nine stations and observed eleven triangles to reach the peak. Two additional stations were included in the triangulation scheme at the midpoint—one on the west bank and one on the east bank of the Torneonjoki. These marked the terminals for the critical baseline.

The next task was to measure the latitude at the northern and southern terminals of the chain of triangles. For this, the expedition was equipped with a 9-foot-tall zenith sector constructed by the London instrument maker George Graham. This large astronomical instrument comprised a long telescope tube suspended at the top end from a vertical wooden beam so that it could swing through a small arc. The amount of swing was measured against a fixed scale and the reference point kept perfectly vertical by use of a heavy plumb bob. To make the necessary measurements, the sector had to be aligned precisely in the meridian plane, the north-south line that passed through the survey station. The entire contraption was suspended within a stout tripod of wooden beams dug firmly into the frozen soil.

Each night from 30 September and 10 October, as the thermometer dropped to $-5°C$, the astronomers took turns in patiently measuring, lying on their backs, the zenith distance of the star delta Draconis as it crossed the meridian of their giant instrument. From a mean of the observations, the difference in latitude between Torneå and Kittisvåra was found to be $0°57'26.93''$.

With the onset of winter, the lakes and rivers of Lapland began to ice over. On the advice of Celsius, Maupertuis proposed to take advantage of the freeze and use the flat expanse of the ice-covered Torneonjoki for his baseline. Accordingly, just before Christmas 1736, the French astronomers and their Swedish colleague, together with Colonel Rietz's men, stepped onto the great ice river and began to set up the delicate measuring equipment. To make the critical measurement, the expedition was furnished with eight precision-made rods each 30 *pieds* (French feet) long. The party then split into two groups to independently measure the baseline distance across the ice.

The cold was intense. Their thermometers would have quickly become useless had they not been kept in glasses of brandy, where the alcohol prevented freezing. The sun barely scraped the horizon at noon before disappearing from view to leave a magical twilit world where the aurora danced its slow rainbow ballet. Except when the snow clouds rolled down from the north, it was seldom totally dark. A curious, delicate silvery luminescence pervaded the scene—starlight and moonlight magnified by a landscape draped in a thick carpet of reflective snow. Despite the vicious cold, taking measurements across the frozen river was

Graham's zenith sector. From *Mesure de Dégré du Meridien en France,* Mémoires de l'Academie Royale des Sciences, Paris, 1736.

an enchanting, haunting experience for men accustomed to warmer, less glacial conditions; moreover, when the two parties compared their measurements, the difference was just 4 *pouces* (4 in., 12 cm) over approximately 8.9 miles (14.3 km).

The baseline measure complete, the scientific party and its military escort returned south to the relative comforts of Torneån wooden houses. The scientists and their assistants were billeted around the little town, and they quickly made friends with everyone, from the butcher to the mayor. Conversation in Suomi was impossible, and young Hellant's services were much in demand; Colonel Rietz also spoke some French. The more prosperous and dignified members of Torneån society spoke Swedish, and Celsius was able to assist with translations. However, when Maupertuis met with the priest of Yli-Torneå, their conversation was conducted entirely in Latin.

Apart from the pleasant social activities offered by the hospitable citizens of Torneå, there was much work to do. Maupertuis and his colleagues had to reduce and adjust all their observations and measurements. From his analysis of the results, Maupertuis calculated that the distance in the meridian from Torneå to Kittisvåra was 54,942.5 toises. But, from the spread of errors in the star observations, he harbored a doubt about the quality of the latitudes they had measured and determined that they would reobserve them as soon as the weather permitted. This was done in March and differed from the previous measurement by 3.49″ (the equivalent of approximately 350 ft, or 106 m, over the ground).

The final angular difference in latitude between the terminal points was averaged as 0°57′28.7″, which when proportioned against the measured distance, gave the length of 1 degree of latitude at 66° 20′ north being the equivalent of 57,437.9 toises (approximately 69 miles, or 104.4 km). Maupertuis's degree was 378 toises longer than Jean Picard's value of 57,060 toises near Paris and "therefore our degree, with aberration, differs by 950 toises from that derived from the measurements of Cassini, listed in his book *Size and Figure of the Earth,* . . . one sees that the earth is consderably flattened toward the poles."[3] The result proved wrong Cassini's prediction that the length would be 950 toises short and vindicated Newton's prediction that, if the earth were oblate, the lengths of degrees of latitude as they got nearer to the poles would increasingly lengthen.

After taking yet more gravity observations with the pendulum, the expedition to Lapland was at an end. Celsius and the French academicians bid farewell to Colonel Rietz and the friends they had made in Torneå and in June boarded their ship, bound for Stockholm. A few days were spent in the Swedish capital, entertained by the local nobility and the king of Sweden himself. Traveling oversea by way of Dunkerque, Maupertuis and his party finally arrived in Paris to present themselves and their results to Cardinal Fleury at the palace at Versailles on 21 August 1737.

Maupertuis's fame was to spread far and wide, but today he is best remembered for his famous principle of least action (1744), from which he believed that the laws of the universe could be unified and the existence of God proven.

The principle is without fault, arguing, for example, that a runner can get from one side of a mountain to the other quicker and with less effort by running around its base than by following the direct line across its summit. As a means for proving the existence or not of God, it had its detractors.

In time, Maupertuis accepted an invitation from King Frederick II (the Great) of Prussia to join his collection of literati at the palace at Sans Souci near Potsdam. In the second year of the War of the Austrian Succession, Maupertuis rather foolishly decided to accompany the Prussian army into battle in order to observe their artillery skills. For his pains, he briefly became a prisoner of war. Fortunately, his incarceration did not last long, and in the year after his release, he was elected a member of the Berlin Academy of Sciences and even held the post of president for 8 years. When he died in the arms of Bernoulli in 1759, King Frederick wrote, *Maupertuis cher Maupertuis, Que notre vie est peu de chose.*[4] Samuel Johnson's friend and sparring partner, the literary critic Charles Burney, said of him, "Would he had been a Christian! I cannot help earnestly venturing to hope that he is one now."[5]

# 5

## EXTREME SCIENCE

ON 16 MAY 1735, THE EXPEDITION OF 31-year-old Louis Godin had set off for Spanish Peru (now Ecuador) to measure the length of a degree of meridian arc at the equator—a whole year before Maupertuis left for the chilly climes and frozen rivers of Lapland.[1] The expedition's two chief scientists were the amiable, aristocratic-looking Charles-Marie de la Condamine (1701–1774), an ex-soldier and noted mathematician, and the ingenious and very talented Pierre Bouguer (1698–1758), the royal professor of hydrography.

It took the equatorial expedition nearly 7 months to reach Cartagena, Columbia. As Peru was then under Spanish dominion, two Spanish naval officers, Jorge Juan y Santacilla (1713–1773), an experienced navigator and naval commander, and Antonio de Ulloa (1716–1795) were appointed by the governor to accompany the expedition as representatives of the Council of Indians. From Cartagena, Godin followed the coast southwest toward Panama, from where he led the expedition through 40 or 50 miles of tropical rainforest to the Pacific Coast. From there, a ship was chartered to take the expedition south to Guayaquil, a sweltering, fever-ridden port set among the swampy islands at the mouth of the Babahoja. It was 25 March 1736.

After a well-earned rest and replenishment of stores, the team set out for its final destination, Villa de San Francisco de Quito, a Spanish-Indian city perched beneath the slopes of the dormant volcano of La Pichincha. From Guayaquil, the men moved north along the coast, surveying the land as they went, until they reached the equator itself. It was decided that to undertake the hazardous route east into the mountains to Quito, the expedition would break up into independent parties. This, it was argued, would allow them to better explore the country and ensure a reasonable chance that some of them might actually survive the trip. These brave adventurers were under no illusions about the dangers they faced.

Penetrating the interior required nearly two and a half months of almost unbelievable hardship. The independent groups with their Indian guides, little men covered from head to foot in red paint, and locally recruited Spanish-Indian porters, to carry the copious stores and precious instruments, traversed treacherous swamps, crept along frightening precipices, and crossed raging mountain torrents. They cut their way through dense, steaming jungle and thick groves of native rubber plants while enduring the extreme heat and humidity, not to mention the persistent attention of the stinging insects and the venomous snakes of the equatorial tropics.

Following the valley of the Guaillabamba River, the route into the mountains became steeper and more rugged as the men climbed higher and higher into the foothills of the Andes. At last, they reached the high valley of Quito; miraculously, they had all survived, and by 10 June 1736 were once more reunited. All about them were the towering, snow-clad peaks of the majestic Andes Mountains. Across the valley, 30 miles to the south, stood majestic Cotopaxi (19,343 ft, 5,896 m), the lofty volcanic cone wreathed in thick white vapors. At the other end of the world, their colleagues had just arrived in the civilized city of Stockholm.

In the high Andes, any difficulties experienced by Maupertuis and his team on the icy Torneonjoki paled into trivial insignificance. The first task was to set out a baseline from which to start the series of triangles that would stretch over 200 miles and cover nearly 3 degrees of latitude. Two months were spent examining the ground and choosing suitable locations for the terminal points. The first baseline began at Yarouqui, a small Indian village nestling in the wide, high Quito valley, where a swath of ground 8 miles long was cleared of obstructions and vegetation.

Using copper-tipped wooden rods each 20 pieds long, the scientists divided into two parties to measure the long baseline independently. After several weeks, laboring in the high-altitude air, the scientists determined that the length of their baseline was 6,273 toises (7.6 miles, 12.2 km). The two independent measurements differed by just 3 inches (7.5 cm), an error of only 0.4 inch per mile: a staggeringly accurate result.

With the first baseline complete, the really tough work began. The apexes of the chain of triangles that would provide the critical distance for the degree of latitude were to be the peaks of the Andean Mountains. At each apex, the horizontal angles had to be measured by using large brass quadrants up to 6 feet across and mounted on inconvenient and heavy cast-iron stands. The weighty and very delicate instruments were hauled on the backs of draft animals or human porters to the great peaks. Once at the top, the surveyors would make camp, then endure the freezing, inhospitable conditions until the weather

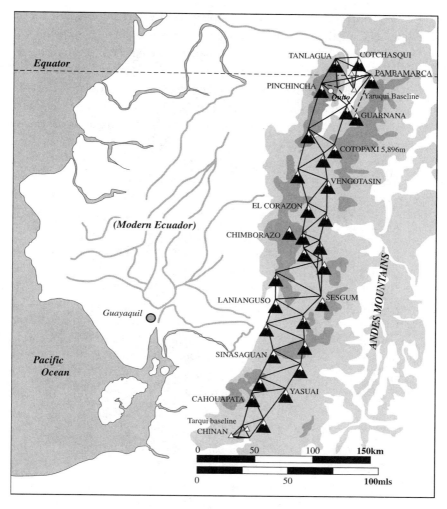

The meridian arc of Peru.

cleared sufficiently to allow an observation; a single angle could take a month to complete.

As the tedious and dangerous business of measuring the mountaintop angles went forward, 200 miles to the south at Tarqui, near Cuença, a 5,259-toise secondary baseline was set out and measured. The method was the same as at Yarouqui, except that a part of the line ran through a large pond and the scientists were obliged to float the measuring rods on its surface.

Because there were two baselines and two methods of computation, there were, obviously, four different conclusions. There were also four different computers. Bouguer, Godin, La Condamine (supposedly a less reliable mathematician than Bouguer), and their Spanish colleagues. Each independently computed the length of the chain of triangles. Godin made two calculations that varied by some 20 toises. The difference between Bouguer's and La Condamine's measurements was 3 toises,[2] a mere 19 feet (5.8 m), but there was an almighty row—La Condamine accused Bouguer of refusing him his right to check his, Bouguer's, calculations. In the event, both men made independent calculations, and some years later the average of the two results, 56,753 toises[3] (110.7 km), was accepted as the length of 1 degree of latitude on the equator.

The observations for the chain of triangles linking the two baselines, together with a host of other important experiments, kept the scientists busy in Peru for nearly 7 years. In addition to measurements of long distances, many experiments were conducted and interesting, groundbreaking discoveries made. Of the more notable were La Condamine's pendulum measurements for gravity and the high-altitude barometric experiments that led to the formulation of the laws relating atmospheric pressure with elevation. On the lower slopes of La Pichincha, La Condamine came across the Quechua kina, or Cinchona tree, whose tough brown bark held promise of a preventative for the terrible ague fever we know today as malaria. He drew a sketch of the miraculous quinine tree and recorded its details, the first European to do so.

In 1741, Louis Godin fell in love with and married Isabella, the 13-year-old daughter of the Spanish governor, and left La Condamine and Bouguer to work on alone. The story of Isabella and Louis is one of the most touching, courageous, and romantic stories of eighteenth-century love. (Alas, this is no place to relate it save that, in the end, after 20 years of separation by the cruel Amazon jungles, Isabella and her lover were reunited.)

Of the many experiments and discoveries made in Peru, two were particularly germane to the theory of gravity and scientific mapmaking. Conducted by Pierre Bouguer, the experiments were the first attempt to detect whether or not Newton's theory (that a large mountainous mass could generate sufficient detectable gravity to deflect a plumb bob) was correct.

In the first experiment, a "second's pendulum" was used to determine the difference in gravity between Isla de las Incas, at sea level, and Quito, on the high tableland, some 9,400 feet (2,860 m) above sea level. According to Newton's inverse square law of gravitational attraction, and if there was no material between the two sites (which clearly there was), the second's pendulum should have shown a difference of 1/1118 in the strength of gravity at Quito than at sea level. In fact, Bouguer found that the difference was 1/1331 less than at sea level. The difference had to be due to the intervening rocky material and the gravita-

Chimborazo, site of the famous gravity experiment.

tional attraction of the high plateau. In time, the relationship became enshrined in Bougeur's law.

For a suitable location for the second experiment, Pierre Bouguer selected
the lofty volcanic peak of Chimborazo (20,560 ft, 6,267 m). Below the precipitous flanks of the mountain, he set up two observing stations more or less at the
same latitude. His two stations and the peak of the mountain formed an oblique
triangle. The station nearest the center of the mountain was just over 2 miles
from the peak, and the more distance station, where he had pitched his tent,
was some 5.5 miles away. Bouguer measured the apparent latitude at each station with the expedition's 12-foot-tall zenith sector with the utmost care. The
ground distance between the stations was then calculated from a simple baseline and trigonometry.

Bouguer used the value for 1 degree of latitude that he had previously determined and, by proportions, applied this distance to the apparent difference in
latitude between the two stations to arrive at a derived "ground" distance. By
comparing this derived distance against the true measured ground distance,
Newton's attractive effect should have been apparent. And it was: the calculation
showed a difference of 7.5 seconds of arc, equivalent to about 750 feet (229 m)

on the ground, but it was much less than the 103 seconds of arc Bouguer antici-pated that such a mountain the size of Chimborazo should have produced. From the results of the experiment, he concluded that the earth's density was some thirteen times that of water—an answer that was clearly far too great.

However, Bouguer was not particularly satisfied with the reliability of his work, claiming, magnanimously, that his answer was so close to the accuracy limit of his instruments that it could not be relied upon to prove or disprove the theory for the attraction of mountains. Nevertheless, the important thing was that he had shown the effect might well exist and that an experiment using better instruments could be designed to measure it. Since Bouguer's day, precise geodetic measurements of the deflection of the vertical have, over the years, shown that the attraction of mountains is indeed less than might at first be ex-pected from Newton's laws. The discovery of the cause for this anomaly re-quired even more exotic experiments and another 100 years.

In 1743, the Peruvian expedition broke up. Bouguer returned to France and was the first to arrive home. La Condamine, on the other hand, crossed the Andes to the headwaters of the unexplored Amazon River, from where he rafted all the way to the Atlantic Ocean. It was an incredible journey of nearly 5 months, which, miraculously, he survived, and he returned to Paris in February 1745. In the honorable tradition of science triumphing over politics, Charles-Marie de la Condamine and Pierre Bouguer were elected fellows of the Royal So-ciety of London.

The two Spanish officers took ships for Europe, sailing around treacherous Cape Horn. Antonio de Ulloa was in a ship that was attacked and captured by the British, and he spent time as a prisoner of war, first in Louisburg, at the mouth of the chilly Saint Lawrence, then in England. He finally returned to his native Spain in 1746. Jorge Juan y Santacilla fared better and on his return to Spain became closely associated with the improvements then taking place in the Spanish navy, writing extensively on the principles of navigation.

In the space of 74 years, the savants of France had achieved the three vital statistics necessary to define a preliminary shape and size for the earth. In the north, Maupertuis had determined that 1 degree of latitude on the ground was the equivalent of 57,437.9 toises. Near Paris, Picard had shown it to be 57,060 toises, and the Peruvian expedition had measured it as 56,753 toises. From this information, it was possible for the academicians to calculate the parameters for drawing an ellipse. This ellipse, when rotated about its polar (north-south) axis formed a spheroid, the shape that best fitted the earth's apparent dimensions.

Newton and his acolytes were proved correct, yet despite the valiant and courageous efforts and extreme science of their academic colleagues, the Cassinis remained unconvinced. What none realized was that they had barely scratched the surface of the problem: things were about to get much more perplexing.

Up to this time, all meridian arcs measured by the French academicians had been observed in the northern hemisphere. Newton's postulation was that the shape of the earth, being symmetrical, would be the same south of the equator as it was on the north. However, this had yet to be proven by direct measurement, and to do the job the académie engaged the considerable intellect of the astronomer, and Jesuit priest, l'abbe de La Caillé (1713–1762).

Nicolas Louis de La Caillé was born on 5 March 1713 at Rumigny, near Rheims, the son of a destitute member of the household of the duchess of Vendôme. Through the patronage of the duke of Bourbon, La Caillé studied theology at the Collège de Lisieux in Paris and took holy orders. Despite his Jesuit education, in the Age of Reason it was science that appealed to his extraordinary intellect. In 1738, through the patronage of Jacques Cassini, he gained employment on the French coastal survey then progressing from Nantes to Bayonne. The following year he was engaged to assist in the remeasurement of the French meridian arc, a task that kept him busy for 2 years. His success on these projects and his insightful observations resulted in election to the Académie Royale des Sciences. In 1739 he moved further up the academic ladder by becoming professor of mathematics at the Collège Mazarin (now L'Institut de France), Paris, where he built an astronomical observatory.

La Caillé and other academicians had for some years lobbied the council of the académie to sponsor an expedition to South Africa to observe the positions of southern constellation stars for the académie's catalogue of navigational stars. It was also hoped that observations of the sun from a location far to the south of the equator, coupled with simultaneous observations in Paris, would allow the solar parallax to be measured, which in turn could be used to determine the earth's distance from the sun.

After lengthy debate, the académie agreed to the expedition and La Caillé was given command. Apart from the stellar and solar observations, La Caillé was to make observations of the moon to augment the scant knowledge of the planetoid's complex orbit around the earth. This would be critical information if the lunar distance method for finding the longitude were ever to succeed. The site for the observations was to be the relatively friendly Dutch colonies in South Africa. There, it was agreed, La Caillé would determine the precise latitude and longitude of the Cape of Good Hope (Cap de Bonne Esperance) relative to the Paris Observatory. In this exercise, his observations were also to coincide with similar observations to be taken by the 19-year-old French astronomical prodigy Jérôme Lalande (1732–1807) at the observatory in Berlin. The separation of the observatories, half a world apart, would provide a baseline long enough to give an accurate measurement of the distance to the moon. It was for all these purposes, and not for measuring a meridian arc, that La Caillé was given charge of the expedition: earth measuring would have to fit in with the astronomy.

La Caillé set off for the Cape of Good Hope in 1750 aboard the French East Indiaman *Glorieux* under the command of Captain Jean-Baptiste Nicolas Denis D'Apres de Mannevillette (1707–1780), an impressive name for an impressive man. D'Apres was himself a great marine surveyor who had studied mathematics and geometry at Paris and was the first Frenchman to adopt Hadley's quadrant (the forerunner of the mariner's sextant) and to use "lunar's" for determining the longitude.

On arrival at Cape Town, La Caillé's first task was to ingratiate himself with Hans Tulbagh, the Dutch governor of the province, which he did with considerable charm. With Tulbagh's assistance, La Caillé built a wooden astronomical observatory on Cape Town's Strand Straat and began his long program of observations. In August 1752, toward the end of his second southern winter, he took some time out from astronomy to travel north on a field trip to scout the country and locate the sites for his meridian arc observations.

At Klipfontein, in a barren valley among the mountains north of Piquetberg, he found a farm that he judged to be at an adequate distance for his purpose from the observatory in Cape Town. He then set about the obligatory business of measuring a baseline from which to build up a series of triangles that would cover the required distance of 1.5 degrees of latitude. An 8-mile-long baseline was established across the Darling Flats, about 40 miles due north of Cape Town. This La Caillé measured by using the customary 4-toise-long (approximately 25-ft, or 7.6-m) wooden rods supported on stands. Governor Tulbagh provided La Caillé with a contingent of soldiers, commanded by engineer officer Captain Eduard Muller, to assist and guard the survey party. By all accounts, the French scientist, apart from being one of the most intellectually endowed men of the age, got along very well with everyone he met and "won the love and friendship of all who knew him."[4]

By early spring 1753, La Caillé was ready to start measuring the angles in the chain of triangles linking his observatory in Cape Town to the farm at Klipfontein. His scheme was, by the standards of meridian arcs, a simple one. From the terminals of the baseline, he observed the angles to Capocberg and Riebeek Casteel, high upon the mountains to the east. From Capocberg and Riebeek, he could see south to Cape Town and north to Klipfontein, about 38 miles distant. He had already determined the latitude on Strand Straat, so only the latitude at the northern farmstead was wanting. Bad weather plagued the observations,

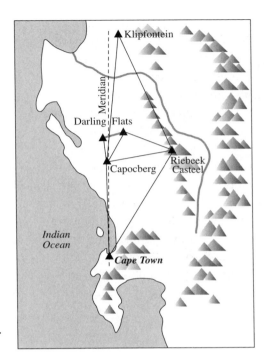

La Caillé's cape arc.

but by the end of the year La Caillé had all the information he needed to calculate the distance from Cape Town to Klipfontein. When he did so, however, he received a nasty shock.

According to his colleague Bouguer's figure of the earth, at the latitude of the cape the distance of 1 degree of latitude should have been 69.16 miles. Instead, La Caillé found from his calculations that his distance was 430 feet (131 m) too short. He checked every observation and every calculation again and again, but there was no mistake: the distance was definitely 430 feet short of what it was supposed to be. If his calculated length of 57,037 toises for a degree of latitude was correct, and he didn't doubt his mathematics, then the earth, instead of being an oblate spheroid flattened at the poles, was distorted and had the form of a prolate spheroid south of the equator (a terrible echo of Jean Cassini and even of Columbus). The earth, according to La Caillé, was pear-shaped.

Although the news was disturbing indeed, there was nothing he could do about it but report the fact. He returned, crestfallen, to his astronomical observations. By the time he completed his work in South Africa, La Caillé had observed over 10,000 new stars, named fourteen new constellations, and could rightly claim to be the most prolific astronomer of all time. On returning to

Paris in 1754, La Caillé was something of a celebrity, a situation he did not enjoy. Distressed, he withdrew into the sanctuary of his Mazarin college to work on his map of the southern stars, *Coelum Australe Stelliferum,* published in 1756, and compile and calculate the thousands of observations he had collected during his sojourn below the equator.

Of the meridian arc, he did very little. It was all too much for him; on 21 March 1762 an attack of gout brought on by the excessive work and long nights laboring by candlelight brought his short but spectacular life to a close. Jérôme Lalande said of La Caillé that, in the short time allotted to him, "he had made more observations and calculations than all the astronomers of his time put together." The quality and quantity of La Caillé's scientific achievements, coupled with his stout Jesuit faith and the "rectitude of his moral character, earned him universal respect" from scientists throughout Europe.

Abbé de La Caillé died never knowing that his observations and calculations for the South African meridian arc were in fact flawless. The reasons for the unwelcome discovery that the earth looked like a pear were soon to be uncovered.

# 6

## ROBBERIES AND DEPREDATIONS

FRENCH DOMINATION OVER THE ART of earth measuring was broken the year following La Caillé's victorious homecoming. The Italians, well known and respected for their assiduous mapmaking exploits, were also interested in the shape of the earth, though for reasons that were more ecclesiastical than naturally inquisitive.

The mid-eighteenth century was an interesting time for the Church of Rome and for Pope Benedict XIV. The Papal Index that condemned as heretical any belief in the Copernican "sun-centered," or heliocentric, model of the universe was still in force. The penalties for those heretics who thought otherwise were also still valid though seldom enforced in the Age of Reason. The church's Aristotelian doctrine, though not its thinking, was at odds with contemporary science. Across Europe, in Catholic and Protestant nations, men of learning had embraced the celestial mechanics of Nicolaus Copernicus (1473–1543) and the planetary laws of Johannes Kepler (1571–1630); few, indeed, remained who thought of the earth as having anything other than a spherical form.

It was in this climate of reason and change that a Jesuit priest, Fra Ruggiero Giuseppe Boscovich, was sanctioned by the pope to measure the first meridian arc in Italy for the Church of Rome. Ruggiero Boscovich (1711–1787) came from Dalmatia, east across the Adriatic Sea, and was a renowned mathematician, astronomer, and natural philosopher. Of formidable intelligence, he had been educated at the college in Ragusa and ordained into the Society of Jesus at Rome (the Jesuits). From an early age he had exhibited a considerable talent for mathematics, a talent that blossomed during his years of study at the Collegio Romano, the most celebrated of the Jesuit colleges. His mentor, Fra Horatio Borgondi, the professor of mathematics, cultivated the young Boscovich's mind and introduced him to the astronomical arts. Boscovich quickly became an accomplished astronomer and was soon widely published on subjects as diverse as sunspots (1736) and the aurora borealis (1738).

Father Ruggiero Boscovich.

As an astronomer and natural philosopher, Boscovich was keen to apply his considerable intellect to advancing knowledge on the shape and size of the earth and to add his own contributions to the growing library of learning. His work and studies would assist the Church of Rome in its cautionary move toward acceptance of the latest science. In time, Boscovich became the church's foremost mathematician, and many universities beyond Italy sought him out to join their ranks. When King John V of Portugal petitioned the general of the Jesuits for ten learned priests to conduct a survey of his dominions in Brazil, Boscovich volunteered. However, the Jesuit's knowledge of hydrology and engineering was as prodigious as his astronomical learning, and instead Boscovich was appointed technical advisor to the papal government. The pope was determined not to lose his valuable science priest and cleverly offered Boscovich the chance of conducting his earth measurements across the plains of the papal states.

In 1755, the 44-year-old rotund and diminutive Father Boscovich, assisted by a fellow Jesuit, Fra Le Maire, began a series of precise meridian arc observations. Two baselines were measured; one stretched along the famous Via Appia south of Rome, and another was set out near the independent principality of Rimini. The priests employed the same principles as the French, using triangulation to extend the baselines to the required distance for the meridian arc, all "accomplished by no slight difficulties"[1] across the heights of the Appenine mountains.

Tha Appian Way, along which Boscovich measured
his baseline.

When added to those of the French expeditions in the northern hemisphere, Boscovich's results further strengthened the argument for an oblate spheroid shape of the earth, at least north of the equator; only God and La Caillé knew what went on to the south. Boscovich published his results, with the blessing of the church and in true ecclesiastical style, as *De litterariâ expeditione per Pontificam ditionem ad dimetiendos duos meridiani gradus et corrigendam mappam geographicam.*

In parallel with the efforts to discover the size and shape of the earth, the sister science of cartography, depicting the earth on paper, was also progressing. Throughout the seventeenth and early eighteenth centuries, the science behind mapmaking had steadily evolved. Better instruments were built, more advanced skills were developed, and the notion of basing a map upon a rigid framework of triangulation was advanced. It is historically convenient to choose some event or point in time to mark the beginning of change. In the history of mapmaking, there are many such transition events that would suffice, but for the earth measurers, the criteria are not the milestones marking advances in the skills of the cartographer or the map compiler or progress in the art of engraving or printing but the measuring process itself.

Mercator's projection had shown how the curved earth could be expressed mathematically on a flat piece of paper. But, to be accurate, to render a view of nature upon a piece of paper, a map, like a house, must be constructed upon a solid foundation. Triangulation had shown that a rigid and accurate network of control points could provide the footings, and now, with a reliable value for the size and shape of the world, the foundations could be planted in the very fabric of the earth itself.

Of the manifold historical events that could be chosen as marking the advent of accurate mapmaking, a windy hill called Sheriffmuir in the Highlands of Scotland offers a solution. Two Scottish armies faced one another across a barren hillside, awaiting the call to battle; it was 13 November 1715. Four thousand royal troops under the joint command of John Campbell, the duke of Argyll, and General John Whetham were arrayed downhill from 10,000 tartan-clad Highland warriors led by John Erskine, the sixth earl of Mar. On a nearby hillock, watching the maneuvers from his horse, sat the famous cateran Rob Roy MacGregor with his brothers and fellow clansmen: he was wresting with his conscience.

Mar's Highland forces moved first, sweeping down in a ferocious "highland charge," yelling and screaming, into Whetham's right flank. It was "a complete rout and prodigious slaughter." Convinced of defeat at the hands of the claymore-wielding, pike-toting wild Highlanders, the timorous Whetham fled the field and took shelter in nearby Stirling Castle. The duke of Argyll was of sterner stuff, and taking the initiative, he cut around Mar's right flank, pushing it back in retreat. The Highlanders may have been superior in number, but Argyll's smaller force of royal troops was better armed. Rob Roy and his small band looked on as the battle unfolded before them across the valley. Watching to see the turn of events, Rob Roy chose not to assist his Jacobite liege lord. Like Stanley at Bosworth Field, he turned about and led his small force away; Rob was not going to add his to the waste of good Scottish blood.

The duke of Argyll's victory at Sheriffmuir marked the end of the second Jacobite rebellion (the first had been in 1688–1689) and the beginning of 30 years of civil unrest. Jacobite conspiracies were not limited to Scotland. Far away from the stormy Highlands, in the south of England, the spa city of Bath was a hotbed

of Jacobite plotting. There were rumors of companies of troops being recruited and stores of arms secretly hoarded. The government decided to act quickly and round up the dissidents before the situation got out of control. Two regiments of dragoons were dispatched to the town, under the command of Major-General George Wade. The rumors turned out to be true; within weeks of his arrival, Wade had uncovered "eleven chests of fire-arms, swords, cartridges, three pieces of cannon, one mortar and molds to cast cannon."[2]

A few weeks later, Wade and his ring of spies exposed a dangerous plot against the body of the king himself, directed by none other than the king of Sweden, Charles XII. Charles had good reasons to detest King George; shortly before his succession, "German George," the elector of Hanover, had allied with Sweden's enemies to seize the bishoprics of Bremen and Verdun. For this reason alone, King Charles, with French sympathies, was a natural supporter of the Jacobite cause and his scheme was to drive King George into banishment and install the exiled Stuart pretender on the English throne. The architect of the daring plot was Count Gyllenborg, the Swedish ambassador to the Court of Saint James in London, whose home Wade raided.

In 1719 came news of yet another insurrection in turbulent Scotland, this time in the western Highlands, when the Jacobites attempted to join forces with the duke of Ormonde. Although it was funded in part by the Catholic Spanish court, the rebellion failed to gain any real momentum and was extinguished by General Wightman at the pass of Glenshiel. But it came as a timely reminder to King George and the Whig government of Lord Stanhope that the Jacobite menace had not gone away. Something needed to be done.

Jacobite scares, outbreaks of violence, and civil unrest continued to plague the British government. Although none was serious enough to warrant concerted action, they reminded everyone of the fragility of King George's claim to the throne of Britain. The spur for action came in 1724 in the form of a memorial to the king penned by Simon Fraser, eleventh Lord Lovat of Castle Dounie, "concerning the State of the Highlands." The Scottish peer warned the king that the Highlanders were very different from the Lowlanders and "do remain to this day much less civilised . . . very ignorant, illiterate, and in constant use of wearing arms, and very expeditious in marching from place to place." Lovat explained to the king the intricacies of the Highland clan system with its "quarrels and jealousies among the chiefs," adding menacingly, "it is no wonder, that the laws establishing the succession of the crown, should be too little regarded by those who have not hitherto been used to a due compliance with any law whatsoever."[3]

Lovat's memorial to the king lamented the "continual robberies and depredations" committed by the wild men of the north. From their impenetrable mountain strongholds, the Highlanders' sudden night raids on the Lowland farms were frequented by "violences and illegal extractions" that the Lowlanders called "black meall," from which we get the modern term "blackmail." The king and his chief minister, Sir Robert Walpole, read Lovat's memorial with interest and alarm, diluted with a dose of healthy skepticism. But King George and his loyal Whigs could not risk ignoring the possibility that there might be some truth in what the Scottish peer wrote. What was needed was an independent assessment by someone that the king could trust—a soldier who could read the land and measure its people. The man King George chose for the job was none other than the man who uncovered the Swedish plot, General George Wade.

King George summoned his general to the palace to discuss Lovat's memorial. On 3 July 1724, Wade received his written royal orders. He was to "inspect the situation of the Highlanders, their manners, customs, and the state of the country in regard to the depredations said to be committed in that part of His Majesty's dominions, to make special enquiry into the allegations that the effect of the last Disarming Act had been to leave the loyal party in the Highlands naked and defenceless at the mercy of the disloyal [i.e., the Jacobites]."[4] Wade set off for Scotland the very next day.

The Highlands of Scotland were a strange and untamed wilderness; an impenetrable fortress of moors and mountains; steep, dangerous valleys; and tumbling streams. The people, too, were wild and violent, at least to their southern neighbors. Their way of life was ancient, untouched by the wider world and ordained by rules wholly alien to the "more civilized" south. It was a land of warriors, living by the warrior's code: blood feuds and unswerving loyalty to the clan chief transcended all other matters, or so it was believed; but, like the shape of the earth, no one was really certain.

Five months later the general was back in the capital, presenting his report to the king. Wade did not allow himself to be overly influenced by the plaintiffs, and the Highlanders' side of the story was reasonably represented. "I proceeded on my Journey," Wade reported, "and have Travelled through the greatest and most uncivilised Parts of the Highlands of Scotland; And humbly beg leave to lay before Your Majesty the following Report, which I have collected as well from my own Observations, with all Faithfulness and Impartiality."

In his report, the general described the land, its boundaries, and its people. Wade had been born and bred in Ireland, and the Highlanders' lot in life was not dissimilar to the impoverished and tortured land of his youth. The Highlanders, he wrote, were structured into various clans, each submissive to its chieftain, and each clan was further subdivided "into Little Branches sprung from the Main Stock who have also Chieftains over them."

General George Wade, by Van
Diest, from J. B. Salmon, *Wade
in Scotland,* 1934.

For defense and policing the Highlands, Wade proposed that new companies of locally recruited troops should be raised and that strongholds and barracks built "for preventing the Highlanders descending in the Low Country in
time of Rebellion," a possibility Wade rated high. In an addendum dated April
1725, Wade made further recommendations: "For repairing the Fortifications of
Edinburgh Castle and Fort-William. For building two New Forts and Barracks at
Inverness and Killihnimen, each sufficient to contain a Batallion of Foot. For
mending the Roads between the Garrisons and Barracks, for the better Communication of his Majesty's Troops." So, in this report of his Scottish travels, we first
learn of General Wade's famous feats of military engineering, constructions
that were to transform Scotland and, in time, even merit the praises of its swift-
footed inhabitants. But, to perform these mighty works, Wade needed maps.

With an illustrious and successful career behind him and the king's blessing
on his report of the state of the Highlands, it was obvious to all concerned that
the 52-year-old General Wade was the man to put his own plans for Scotland
into effect. With the grand title of commander-in-chief of His Britannic Majesties Forces in Northern Britain, General Wade rode into the city of Edinburgh
on 16 June 1725. He arrived well prepared for his duties. Before his appointment,
there had been just four permanent garrisons of loyal troops patrolling the vast
Highland region. Now, six companies of the new regiments and four battalions

of foot soldiers were on their way to the Highland fortress commanding the northern entrance to the Great Glen at Inverness.

The majority of the troops were themselves Highlanders, commanded by Highland officers, and were billeted in new forts and rebuilt redoubts strategically located across the country. These Highland regiments would, in 1739, become more familiarly known as the Black Watch, one of the fiercest, bravest, and most respected regiments of the British army.

On 31 July 1725, 6 weeks after he had arrived in Edinburgh, Wade was able to report to the king that his military preparations were already showing success: "Several of the Chiefs of the Highland Clans sent to me . . . assuring me they would peaceably surrender their arms, pay a dutiful obedience to your Majesty's commands, and a punctual compliance with the Disarming Act."[5] It was an excellent start; in fact Wade would have very little trouble with the clans as far as disarming went. When trouble did come, and it came quickly enough, it had nothing to do with the disarming of ferocious clansmen but arose over the government's enforcement of a universally unpopular malt tax. The tax was an unfair levy on the Scottish maltsters, who were required to pay an extra sixpence on a barrel of ale, and was contrary to the spirit of the Act of Union, which implied that both England and Scotland would enjoy equal treatment in matters of trade.

In handling the malt tax riots, Wade demonstrated an ability to balance force with diplomacy and persuasion. He had the situation under control in less than a month. Once peace and honor had been restored, Wade felt secure enough to leave Edinburgh for his main mission: collecting arms from the clans. In this task Wade suffered no illusions; he expected the disarmament process to be difficult and probably dangerous. Embarking aboard HMS *Rose* on 1 August 1725, he set sail from Edinburgh and headed north along the Scottish east coast toward his Highland stronghold at Inverness. The weather was against him; after 4 days of heavy seas and roaring gales, the *Rose* was forced to put into Stonehaven, and Wade was obliged to complete his journey on foot.

The overland trek to Inverness was an opportunity for him to take a closer look at the country over which he had transient dominion. It was indeed a wild land—wild and beautiful and rugged. Plunging cliffs bordered the sea, and the waves of the North Sea crashed onto the inhospitable shore with "all the terrifick grandeur of the tempestuous ocean."[6] Fifty years later, Samuel Johnson would follow the same path, stopping off with his friend, the 23-year-old James Boswell, to admire the crags and countryside—"still naked, the hedges are of stone." Traveling via Aberdeen and following the coast road, such as it was, Wade arrived in Inverness on 10 August. Here he received a pleasant surprise when he was greeted by some fifty Scottish chiefs, each one ready to pledge allegiance to King George and to peaceably hand over their weaponry.

By the end of his first year in Scotland, Wade was able to report to King George that many of the criminal practices, especially the odious tradition of "black meall . . . is now no longer paid by the inhabitants bordering on the Highlands; and robberies and depredations . . . are less frequent."[7] The Lowland farmers lived with a better prospect of peace than they had for many years. With the peaceable enforcement of the Disarming Act and a drop in the crime rate, together with an amicable settlement over the antagonistic malt tax, General Wade was free to consider the next phase of his "civilisation" of the Scottish Highlands.

In his report to the king, General Wade had noted that the country of the Highlands was almost impenetrable and that the roads, or what few rough tracks passed as such, were in a deplorable condition. His real concern though, was not for easing travel for the "very expeditious marching from place to place" of the inhabitants but to facilitate the expeditious movement of his majesty's troops and cannon.

Wade's famous road-building campaigns began in May 1726 with the opening up of a paved way between Fort William, at the western end of Loch Ness, and the garrison of Fort George, near Inverness at the eastern end of the Great Glen. The laborious work was undertaken entirely by the soldiers, watched from a distance by the disapproving Highlanders.

This feeling of disapproval with the new-fangled roadways was to change when the populace discovered "the ease and convenience of transporting their merchandise" over the road's smooth surface. This, the first of General Wade's military roads, ran as it still does today, along the south shore of Loch Ness. Over the next 13 years, Wade built some 250 miles (400 km) of roads that traversed the Highlands and opened up the interior to ready access from the south. Stone bridges were built to span the many mountain streams and rivers; some were just simple stone vaults, and others, like the elegant bridge across the Tay at Aberfeldy, were sublime structures with soaring arches.

At the time, road building in Britain was something of a novelty. There were few models to follow other than those built by the Romans 1,400 years earlier. Wade's building techniques (he had some experience of military road building during the attack on Minorca) followed the classical military pattern perfected by long-dead legionary engineers. As a generality, wherever possible he built in straight lines and climbed the steeper hills by using zigzag traverses. In this task Major William Caulfeild, a young Irish engineer and son of Viscount Charlemont, ably assisted him.[8] After Wade returned to his military duties in 1739,

Caulfeild stayed on in Scotland as inspector of roads. It was this largely forgotten engineer who extended the communications network of military roads to the east and south and who eventually built more miles of road, before he retired in 1767, than Wade ever did.

The roads were not planned arbitrarily; routes were carefully thought out to link the military strongholds by the most expeditious course, predating Maupertuis's principle of least action, to provide rapid access into the Highlands from the Lowlands and England by troops, horses, and cannons in time of insurrection. It is quite impossible to conceive that, unlike the Romans, Wade planned and built his military roads without the aid of a map. What map, if any, he had to start with we do not know. However, Samuel Smiles made this interesting observation:

> Very little was known of the geography of the Highlands down to the beginning of the seventeenth century. The principal information on the subject being derived from Danish materials. It appears, however, that in 1608, one Timothy Pont, a young man without fortune or patronage, formed the singular resolution of travelling over the whole of Scotland, with the sole view of informing himself as to the geography of the country, and he persevered to the end of his task through every kind of difficulty; exploring "all the islands with the zeal of a missionary, though often pillaged and stript of everything; by the then barbarous inhabitants." The enterprising youth received neither recognition nor reward for his exertions, and he died in obscurity, leaving his maps and papers to his heirs. Fortunately, James I [Scottish heir to the English throne] heard of the existence of Pont's papers, and purchased them for public use. However, they lay unused for a long time in the offices of the Scotch Court of Chancery, until they were at length brought to light by Mr. Robert Gordon, of Straloch, who made them the basis of the first map of Scotland having any pretensions to accuracy that was ever published.[9]

Timothy Pont was, allegedly, a Scottish clergyman, or minister, who began his surveys of Scotland around the year 1583. He never found a publisher for his work, although his map of Lothian and Linlithgow was included in Mercator's *Atlas* edition of 1630. As Smiles attests, Pont's manuscript maps and drawings were acquired by Robert Gordon and his son James, who redrew and corrected the maps and added some work of their own. The maps were finally published in 1654 in all the major European languages (including Latin) except English.

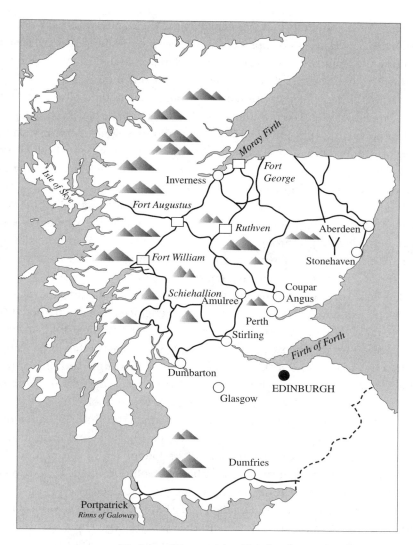

Wade's military roads of Scotland.

Perhaps General Wade had a copy of the Pont-Gordon map on which to plan his campaign of construction; we shall never know. Surveys for the roads were certainly undertaken and examples of the maps and sketches the civilian and military surveyor-engineers produced can be seen in the National Library of Scotland. The cartographer, John Avery, was engaged to compile a map of the military roads and their environs "begun in the Year 1725 and continued and done at times as it was required to the present Year 1730." This map was at a scale of 1 inch to the mile, a scale that was to be the primary one for many maps of the future.

That Wade eventually had a good map of the Highlands is also evidenced by General Henry Hawley's complaint against the Jacobites in 1745 on the eve of entering the Highlands: "I am going in the dark; for Marechal Wade won't let me have his map; he says that his majesty has the only one to fellow it. I could wish it was either copied or printed or that his majesty could please lend it to me."[10] In fact, Hawley was not completely in the dark for he had a copy of a map first compiled in 1731 by John Elphinstone, a military engineer serving with the duke of Cumberland, and probably based on, or including, much of Avery's work. The map was rendered at a scale of 13 miles to the inch, which, Hawley discovered to his cost, was quite inadequate for military maneuvers.[11]

Wade, in his correspondence, frequently refers to "measured miles" between various locations. The names of the surveyors who made these measurements, who found the routes through the mountain passes and around the swamps and made the detailed drawings for the bridge crossings, are lost except for Gordon and Morrison, two "engineers" engaged in 1749.[12]

The Board of Ordnance, responsible for funding and providing for the military works, employed both civilian and military engineers, but the methods they used to make their maps were almost certainly after the military style. An excellent method for making small-scale maps quickly and accurately is by using a Roman invention called a "plane table," a sort of drawing board on a tripod. However, the British army, who put its faith in a compass, never favored this expeditious method until it became the popular surveyors' tool in India.

The means employed by Wade's surveyors to map the Scottish Highlands is conjecture but was probably compass bearing, pacing, and employing another Roman invention called a "perambulator" or "odometer," a contraption that the surveyor used by counting the number of revolutions made by a wheel of known circumference. The details of the land were sketched by eye.

However accurate Wade and Caulfeild may have considered the work of their surveyors, and despite Hawley's lament, the maps produced by Avery were not of a high standard or particularly accurate and at best could be described as route maps. Nevertheless, they were probably the earliest formal military maps of any great extent ever made in Britain, and possibly in Europe, and their value in revealing the topography of a hostile land and their potential for guiding armies through the mountainous terrain was not lost on the British government.

It is ironical, then, that the first to use Wade's roads for the purpose for which they were designed was not His Majesty's loyal troops but the rebel armies of Charles Edward Stuart, the Young Pretender, who found them most convenient for marching his Highland warriors south for his invasion of England.

# 7

# A MAGNIFICENT MILITARY SKETCH

WHEN GENERAL WADE BUILT HIS ROADS, the Highland clansmen still spoke the old Gaelic (Irish) language called Erse and affected the Irish "tartan" manner of dress. Most of the Lowland aristocracy, wealthy burghers and farmers, had adopted many English customs and lifestyles, and the poor were virtually indistinguishable from their English cousins. Highlanders regarded the Lowlanders with the same disdain they apportioned to foreigners; the Lowlanders looked upon Highlanders as barbarians.[1]

One rebellion might have been put down at Sheriffmuir, but the Jacobite cause still smoldered in the background. On a number of occasions during road building, Wade had to investigate reports of Jacobite plots or send out patrols to arrest activists. Some measure of the ire of the Whiggish landed classes of England with the Scottish situation, and their faith in Wade to solve it, can be appreciated from the new verse that was incorporated into Britain's national anthem:

> God grant that Marshall Wade,
> May by thy mighty aid,
> Victory bring.
> May he sedition hush,
> And like a torrent rush,
> Rebellious Scots to Crush,
> God save the King.

Fortunately, Wade was above such sycophanctic nonsense and held the more honorable Scots in greater esteem than he did many Englishmen. Meanwhile, in 1727, King George died and his estranged son was crowned George II. For a short time it seemed as if the new king, who had loathed his father and the

59

Whig cronies who had kept him in power, might dissolve the government and install his own companions. In the event, prudence and the status quo prevailed, and the Whig party continued to enjoy its political supremacy.

In 1740, Britain found itself embroiled in the War of the Austrian Succession, and the Jacobites saw another opportunity to advance their cause. In July 1745, the son of the exiled James, Prince Charles Edward Stuart, landed in Scotland and quickly succeeded in raising a force of some 2,500 loyalists and sympathizers. Two months later he routed the royal army in a pitched battle on the windswept salt fields, called Prestonpans, beside the Firth of Forth. The prince's resounding victory, so early in the campaign, encouraged more Scots and Jacobites to flock to his banner. With his army swollen to 5,000 men, Charles Edward marched confidently down Wade's convenient roads into England with high expectations that the southern Tories and Catholics would rally to the Stuart flag.

In the event, the rebellion foundered from lack of popular support and was brutally extinguished by the duke of Cumberland at Culloden in 1746. Yet the political fallout was not all bad, and in time there came a renewed, more vital interest in Scotland. As the sorry lot of the country became clear to the southern half of the nation, perhaps for the first time since the Act of Union, so also did the depth of the gulf that divided Scotland politically and economically from England.

In Europe, the War of the Austrian Succession rumbled into its sixth bloody year. The British government remained unsettled and suspicious of the loyalties of the northern half of the realm. Peace, it was argued, would be more sustainable if the "country, so very inaccessible by nature, should be thoroughly explored and laid open."[2] A greater effort, as well as interest, was expended into improving and expanding the network of roadways started by Wade. More improved fortifications and garrisons for troops were constructed.

The need for a comprehensive map of the Highlands was plainly evident. None other than the duke of Cumberland had brought the idea of a Scottish map to the attention of his father, King George, probably as a result of seeing Wade's map and the knowledge that the French were busy on their own project.

As well as determining the first reliable estimates for the size and shape of the earth, France was also advanced in the science of surveying and mapmaking. In 1681, Jean Cassini and Jean Picard, with the assistance of many other scientists, had completed a great survey of the coasts of France. The resultant map invoked the displeasure of the king of France, who complained that their work had done more than had any enemy in reducing the size of his realm.

Until Cassini's map, French maps were as poor in accuracy and reliability as those of any nation. Cassini, enthused by the Academié Royale de Science's success with triangulation for meridian arcs, advocated the use of that method to cover all of France with a rigid framework upon which he could base an even more accurate map of the country. The idea received much acclamation, but the project was never begun in his lifetime and it passed to his son, Jacques. In 1733, King Louis XV issued his royal command for la Carte de Cassini, a project that would eventually require a triangulation scheme comprising over 800 individual triangles. In 1745, Jacques Cassini laid his first draft of the map of France before the king.

The duke of Cumberland pointed out to his father and the chief ministers of state the advantages and importance that such maps could have in putting down further rebellion in Scotland and how they could aid with the collection of taxes. The duke was no foreigner to mapmaking, having learned from his military training and during the Flanders campaign that a decent military map was invaluable. In 1746, as he led his army's advance on Aberdeen, the duke invested 6 weeks of precious time sending out engineers and scouting parties to survey the area and prepare an assault map at a useful military scale.

A week before Bonnie Prince Charlie fled Scotland for exile in France, another event occurred on the other side of the world that would ultimately cause a military genius to appreciate the advantages of good maps. A young Robert Clive, together with the entire European population of Madras, was preparing to evacuate the city. Since its foundation in the sixteenth century, the East India Company had managed its overseas estates and possessions as an independent trading firm. But in India, where it was in almost perpetual dispute with a French rival, the burden of proprietorial responsibility was increasing.

Until 1748, the East India Company had never had recourse to call upon the home nation's assistance in managing its more violent negotiations. A small, mostly native army led by European officers was sufficient to maintain its security, or so its London directors believed, and on the whole they had been right. Then, in September 1746, the French commander of the region, La Bourdonnais, with his soldiers and native contingents swept into the outskirts of Madras and besieged Fort George. The company's force was hopelessly outnumbered and, on 10 September, wisely surrendered to the Franco-Indian forces. This single act of aggression was to have serious repercussions and well demonstrated the difficulties faced by military gentlemen when trying to navigate their men, cannon, baggage trains, and camp followers through uncharted lands. In this ignominious defeat, Clive learned a very useful lesson that he would not forget: a map can save your life.

However, it was along the rocky shores of Loch Ness, at Fort Augustus, that mapping history was made. In 1746, the fort was being rebuilt after its destruction during the Jacobite uprising. It was the duke of Cumberland's headquarters and was then under the command of Major-General Lord Blakeney. Also using the camp as his headquarters was the deputy quarter-master general of North Britain, General David Watson (1713?–1761). Watson was a native Scot and an engineer by profession and had been in charge of one of Caulfeild's road-building parties before his promotion. His experience on the roads and his knowledge of the country had made him acutely aware of the disadvantages the military suffered from its lack of maps.

Watson, who was at the battle of Falkirk and with the duke of Cumberland at Culloden, shared his royal master's concern over the deficiencies of their maps when they "found themselves greatly embarrassed for want of a proper Survey of the Country."[3] History has it that it was General Watson who "first conceived the idea of making a map of the Highlands."[4] With the duke's encouragement, Watson sent his proposal for a "geometrical survey . . . executed at the public expense"[5] to the government in London. The wisdom of the design, the efficiency with which Watson claimed it could be executed, and the support of the king and his powerful son meant that approval was prompt.

The work of the quarter-master general's office, within the auspices of the Board of Ordnance, was principally concerned with the building and maintenance of defensive works and the upkeep of the military roads. In this regard, Watson's staff was carrying on the work started by Wade over 20 years before, work that, under William Caulfeild, was beginning to radically change the nature of travel in Scotland. Watson, who received an additional 5 shillings a day for his post, had two assistants, "Stewart and Mr Roy who have 4s a day each."[6] Four shillings a day for a young civilian Scot in 1747 was a great deal of money, and it can be concluded that these two assistants were probably Watson's "works superintendents."

In July, the British government approved the Board of Ordnance to fund Watson's ambitious proposals for a map of the Highlands. Watson determined to start the work immediately, and he turned to his assistant William Roy (1726–1790), a 21-year old draftsman who would one day become one of the world's most celebrated scientific mapmakers and earth measurer.

William Roy was born on 4 May 1726 at Milton Head, Carluke, in Lanarkshire, first son of John Roy, factor to Sir William Gordon of Milton, and his wife, Mary Stewart. A factor, or land agent, to a great landowner was not a poor man, and young Roy enjoyed a lifestyle considerably better than did his local contemporaries. He was sent to school in Carluke, a small market town in the central Lowlands of Scotland, and completed his education at the grammar school in Lanark. Unlike his younger brother James, who attended Glasgow University

General William Roy, a cartoon, the only known image of him.

before entering the church, William Roy left school for a minor position in the post office at Edinburgh.[7]

In 1747, Roy appeared on Lieutenant Colonel Watson's civilian staff at Fort Augustus. He did not begin his army life as an officer but as a civilian. Tradition has it that Roy's career began as a surveyor of the post roads for the Edinburgh Post Office. If so, this job could not have lasted long for almost certainly he was a draftsman for the Board of Ordnance at Edinburgh Castle shortly before he was engaged as one of Watson's assistants.

Roy's mode of entry into the officer corps is not known with certainty, and it was not until 23 December 1755 that he was gazetted a "practitioner engineer." Roy was familiar with Wade and Caulfeild's maps of the area, and in David Watson he had the benefit of a clever and capable tutor. The young man's mathematical skills were adequate (in later life they would reach very high levels of accomplishment), and all accounts of him speak of a lively, inquisitive, and assiduous character. To some, his ardor must have been galling; to others, it drew the highest praise. A tall and gangling youth, William Roy was a quick learner, more comfortable with clever persons than with the less mentally agile, and just a little dour.

Beginning in 1747, the survey work went ahead "under the auspices of the Duke of Cumberland"[8] in his capacity as master-general of the Board of Ordnance. For the first 18 months, the young Roy worked alone on the project. In 1748 "the Specimens of his [Roy's] progress were so satisfactory, that it was determined to extend the Survey over the whole of the North of Scotland."[9] The next year additional resources were recruited, and the work was extended further afield. At its peak, the Scottish survey had seven survey teams, each with a civilian surveyor or engineer officer, a noncommissioned officer, and six private soldiers who, as did their predecessors who worked for Wade, received additional pay.

Military practitioner engineers, John Manson, Hugh Debbeig, and Lieutenant Williams, led three of the teams. The other crews were led by civilians—Roy; David Dundas (Watson's nephew, later General Sir David Dundas), who became a lifelong friend; William Dundas (who also become a general); and Howse (later the Reverend Howse). The map drawing was primarily under the direction of the Board of Ordnance draftsman Paul Sandby (1725–1809), "the father of English water colour art,"[10] whose delicate skills and artistry greatly influenced the young Roy.

To some extent, the peace treaty of Aix-la-Chapelle in 1748, ending the War of the Austrian Succession, had eased the strain on the government's finances and released men who might otherwise have found themselves sighting along a musket barrel and not peering through a surveying instrument. In true military fashion, the Scottish survey was accomplished by compass, circumferentor,* and perambulator, or the primitive engineers' chain. The details of the map, the roads and rivers, were "paced," whereas the hills and mountains were merely sketched roughly in position. The surveying work occupied the temperate season, but the drawing of the maps was done in the drawing room at Edinburgh Castle during the winter months.

The progress and results of the Highland survey were so successful that the work was extended to include the Scottish Lowlands. The final map, at a scale of 1,000 yards to the inch (1:36,000), comprised eighty-four rolls of varying size but was only ever produced in manuscript form.[11] For orientation, the maps were drawn with their sides aligned to magnetic north, which, at the time, was about 17 degrees west of true north. Paul Sandby returned to his duties at the Tower of London in 1751 but was recalled 2 years later to render the Scottish maps to a smaller scale. This he did "with a thousand graces" at a scale of 4,000 yards to the inch (1:144,000) and aligned the new map to true north.

*Also known as the "common theodolite," it was a circular device with a cruciform of four sights.

Detail from William Roy's map of Scotland. Courtesy Ordnance Survey.
© Crown copyright.

Britain's relations with France had been buckling under political strains ever since the conclusion of the War of the Austrian Succession. But it was in America, far away on the other side of the ocean, that the greater conflict erupted when an obscure Virginian soldier by the name of Colonel George Washington killed nine French patrolmen during a fracas in the wilderness of Ohio. Thus, the Scottish survey came to an abrupt end in 1755 because the men, as Roy put it, were "furnishing service of other kind."[12]

In America, the resulting conflict is remembered as the French and Indian War, a time when Britain and France grappled for dominion of the vast North

American wilderness. The following year, 1756, the European war exploded. Britain, allied with Prussia and Hanover, faced the armies of France, Russia, Austria, Saxony, and Sweden. Known as the Seven Years' War, the conflict was the first that spanned the Atlantic Ocean, and it could justly claim to be the first "world war." It was to prove a pivotal period in European history, especially for Great Britain. The conclusion of the war by the Treaty of Paris in 1763 would see Britain's total dominion over North America east of the Mississippi, much of India, and the islands of the West Indies. It marked the beginning of Britain's great commercial empire.

On 23 December 1755, the 29-year-old William Roy was posted practitioner engineer and commissioned into the 53rd Regiment of Foot. His work on the Scottish survey was widely acclaimed, and he soon found himself employed on reconnaissance duties around Britain, preparing maps and auditing defensive positions against a possible invasion by France. Along with David Watson, Roy was transferred to duties in southern England, where attacks were most expected.

In June 1756, he surveyed a strip of country between the towns of Salisbury and Dorchester and produced a map at a scale of 1 mile to the inch (1:63,360). In May 1757, he was commissioned as an ensign in the newly created Corps of Engineers[13] and posted to the army in France. In 1759 he was again promoted, this time to captain of engineers, and took part in the Anglo-Hanoverian assault against the French forces in Minden, Germany, where he drew a map of the battle.

Roy's experience under Quartermaster General Watson led him to be appointed deputy quartermaster general of British forces in Germany, a position he held from 1760 to 1761. In 1762, he was made a lieutenant general in the army. The Scottish map had been laid aside; it was not far from being completed, but even at that early date Roy recognized that it was by no means perfect. "Although the work . . . possessed considerable merit," he reminisced in 1785, "and perfectly answered the purpose for which it was originally intended; yet having been carried out with instruments of the common, or even inferior kind, and the sum allowed for it being inadequate to the execution of so great a design in the best manner, it is rather to be considered as a magnificent military sketch, than an accurate map of a country."[14]

Roy's "magnificent military sketch" was, nevertheless, a detailed map in the modern sense of the word. It showed roads, rivers, settlements and towns (often in great detail), hills, mountains, and passes. The artist Paul Sandby had developed recognizably modern cartographic methods for depicting the natural features of the landscape. For example, techniques such as hachuring and hill shading were used to give the impression of the topographical relief of the land.

This map set the tenor for British military mapping for the next half century. Its influence can be seen in Charles Vallancey's survey of Ireland (1778–1790), in the maps of the Saint Lawrence produced by George Murray in 1761, and in the maps of the American East Coast charted by Cook's friend Samuel Holland between 1764 and 1775. The style and content of the Scottish map is also reflected in the surveys of Bengal produced by James Rennell, of which more later. Of more importance, Roy understood that what the whole enterprise lacked was the sort of rigid underlying geometrical framework of Cassini's map of France— a foundation made up from triangulation and precise latitudes and longitudes. He determined to teach himself higher mathematics, master the astronomical sciences, and learn more about the shape of the earth.

# 8

## PERSONS WELL VERSED

WHEREAS THE BRITISH LAGGED BEHIND the French in the science of geodesy and mapmaking, they were miles ahead in that of astronomy, especially for navigation. Despite setbacks, Britain's overseas trade had been expanding steadily and was becoming increasingly vital to the country's economy and prestige.

Britain's position as the world's supreme maritime power was also becoming evident, especially to its enemies. Navigating the treacherous oceans with some degree of certainty and protecting valuable cargoes, not to say ships' crews, from shipwreck demanded accurate and reliable charts, which in turn demanded reliable navigation. To this end and following the example of France, King Charles II had the royal architect, Sir Christopher Wren, design and build Britain's first modern observatory.

Completed in 1675 and funded partly from the sale of old gunpowder and unwanted military stores, the elegant observatory was constructed on the top of a bluff overlooking the River Thames at Greenwich. The sole purpose of the Royal Observatory was to provide accurate tables of the positions of the stars and planets for the use by navigators in finding their position at sea. To make the necessary observations, the great astronomer John Flamsteed was installed as England's first astronomer royal.

In 1759, the incumbent astronomer royal was the Oxford don James Bradley (1693–1762). His only assistant, or "labourer," as he was known, was Charles Mason (1728–1786). It was a hard, thankless, and lonely existence to be closeted in the draughty observatory out on the heath, far from the polluting smog of London's air. Both men were engaged in the observations, but as the elderly Bradley's health began to fail, the nightly ritual fell increasingly to the younger man. An alarm clock summoned Mason to work at all hours of the night, often all night, and his day was spent on reducing the observations and calculating the results. Between them, these two astronomers observed and catalogued the position of hundreds of "navigation" stars, as well as recording precisely the po-

Royal Observatory, Greenwich, by Thomas Shepherd.

sition of the moon in its monthly journey across the sky. Eclipses of the sun and moon were diligently recorded, as were the appearances of comets, eclipses of the satellites of Jupiter and Saturn, and many other valuable stellar phenomena. Bradley and Mason had also pinned down the latitude of the observatory's most important instrument, the transit circle, to 51°28′40″ north, "probably without the error of a single second."[1] The longitude, being arbitrary was, of course, 00°00′00″.

Charles Mason was born in April 1728, the third son of Charles Mason, the village baker and miller of Oakridge Lynch, Gloucestershire. Brought up in the then prosperous wool lands of the western Cotswold hills, the Mason family was moderately well off, even by the standards of the affluent working people of the area. They lived in a pleasant, mellow limestone cottage, tiled with the same stone and boasting a lintel above their front door robbed from a local abandoned monastery. Like many cottages in the area, Wear Farm was leased from the bursars of Brasenose College, Oxford.

The Masons were devout Anglicans and worshipped in the old church of St. Kenhelm, recently renovated in the Queen Ann style, in the nearby parish of Sapperton. Their view of the world conformed exactly to that of the established Church of England and King James's version of the Holy Bible. The world, as everyone knew, had been created in 6 days and, thanks to the learned theological chronology of James Ussher, at the date of baby Charles's birth was 5,732 years old. No longer did any of the more learned in the village believe that the tropics were zones wreathed in fire or that Jerusalem lay at the center of Creation. Monsters there certainly were, though no one could say precisely where, and barbarians and cannibals unquestionably populated foreign lands.

Charles grew up, a normal, boisterous village lad, in a rural society that was certain in its beliefs and of its place in the world. It was a confident place, where the villagers and swains might acknowledge the superiority of the squire but were, almost, free of his rule. Yet it was not a constant world that the young Charles inherited. Long-held beliefs and Holy Writ were being challenged, albeit cautiously, by men in the wider world, and dissenters were shaking the very fabric of Anglicanism.

A rudimentary village education, much dependent upon the family's Bible, was followed by study at Tetbury grammar school, which was too far from his home for daily travel. Charles boarded in the town, possibly with the headmaster, Robert Williams, who was to become a great friend and mentor to the young man. Williams was a keen amateur astronomer and collected all manner of scientific instruments. He must have been delighted with his new young student when Mason displayed exceptional abilities for geometry and mathematics. During his school days or soon after, Charles Mason received extra tuition from a noted mathematician by the name of Robert Stratford.

The 10 years of Mason's life after completing his schooling is a mystery; he may have earned a living as a teacher or helped with the family business. Through Stratford or possibly Williams, Mason's singular abilities eventually came to the attention of the Astronomer Royal and Oxford don James Bradley, whose country house at Chalford was just a few miles from Mason's more humble dwelling. In 1756, Bradley offered Mason the job of live-in assistant at the Royal Observatory. The salary was mediocre, a mere £26 a year, but there were perks from showing visitors around the observatory.

About the same time as the job offer, and probably facilitated by it, Charles married Rebekah. The newlyweds moved into the assistant's apartments at the observatory, a loft sandwiched between the transit room and the quadrant room. Their first child, William, was born in 1757, and another son, Isaac, arrived in 1758. One wonders how the sounds and inconveniences of family life, with young, noisy babies, affected the solemn business of astronomical observation. Bradley was an aging bachelor, but perhaps he relished the idea of a young, busy family invading his cloistered life. We shall never know.

Occasionally other astronomers would visit the observatory and assist with the observations, especially if there was something interesting to study. Their journey from filthy, smoky London to stately Greenwich could be one of great

Nevil Maskelyne. Courtesy
Smithsonian Institute.

tranquility, as James Boswell recorded in 1763 when he and Samuel Johnson "walked to Billingsgate, where we took oars and moved smoothly along the silver Thames. It was a very fine day. We were entertained with the immense number and variety of ships that were at anchor, and with the beautiful country on each side of the river."[2]

One of the observatory's most regular and, to this story, most important visitors, was a young Cambridge fellow and ordained cleric by the name of Nevil Maskelyne. Born in London on 6 October 1732, Nevil was the third son of Edmund Maskelyne of Purton, Wiltshire. His father had died when Nevil was only 12, and the family was left rather poor by the standards of their class. He was educated at Westminster School in London, where his contemporaries included the poet William Cowper, as well as many other boys who would later achieve fame, among whom were Warren Hastings, first governor-general of India; Charles Churchill; George Colman; and the earl of Dartmouth.

In July 1748, shortly before the death of his mother, Elizabeth, Nevil observed an eclipse of the sun—an event that stimulated a lifelong love of astronomy just as it had for Jean Picard 100 years before. In 1749 Nevil went to Cambridge University (the family members were squirearchy and Whigs), first at Catharine Hall, then Trinity College, from where he graduated in 1754, the year before his sister Margaret married the future Lord Robert Clive of Plassey (Clive of India). A bachelor's degree in divinity, mandatory for any prospective Trinity fellow, was taken in 1757, and Nevil was elected a fellow of his college in the same year. It was about this time that Maskelyne was introduced to the astronomer royal and his assistant, Charles Mason. The next year Maskelyne—"a person well versed in mathematical learning and natural philosophy"[3]—was elected fellow of the Royal Society.

In 1759 Britain assumed its imperial mantle. The great victories at Quebec and in Quiberon Bay on the west coast of mainland France were buoyed by the news of lesser victories as far afield as the Caribbean and India. But, for the 31-year-old Mason, it was a very sad year; in February, Rebekah died, leaving him with the two young boys to bring up alone. Sad as it was for Mason, it was an important year for British mapmaking. The Society for the Encouragement of Arts, Manufactures and Commerce (now the Royal Society of Arts), announced an award of £100 for the best surveys of any English county conducted at a uniform scale of 1 mile to the inch.[4]

Inspired, perhaps, by Henry Beighton's map of Warwickshire and the cartographic work of the Cassinis in France, the society's initiative, backed by the government who wanted maps but at no cost to the Exchequer, was to stimulate a surge of mapmaking. By the end of the century, more than thirty counties had been mapped. Twelve of the mapmakers were rewarded by the society for the excellence and quality of their work. In the years to come, these maps were to play an important role in the development of the British national survey.

It would be wrong to suppose that all this mapmaking was something new to Britain. The latest rash was just one of many tangible examples of the confidence Britain was beginning to experience, perhaps for the first time since the Tudors and despite suffering severe setbacks in the ongoing war with France. Britons were beginning to realize their great prospects, as well as the great profits, from worldwide trading, protected in turn by the world's most powerful navy. There were also encouraging signs from India that the East India Company's worst days were behind it.

In the previous century, it had been the Dutch who supplied the world with its maps and charts,[5] but, after 1750, increasingly it was London publishers and engravers that were taking the lead. These new "cartographers," as they were called, would set the standards for mapmaking in the years to come and heavily influence the future styles that we now take for granted. Perhaps one of the most influential of the time was Thomas Jefferys (1695–1771).

Jefferys was an outstanding cartographer and one of the leading producers of quality maps of the Americas. He was also geographer to the prince of Wales and King George III. By 1759, Jefferys had already produced at least one county map at a scale of 1 inch to the mile. In 1749 he published his *Shires of England &*

*Wales,* a reissue of Saxton's *Atlas of England & Wales,* and in the same year (with Thomas Kitchin) the *Small English Atlas.* In 1755 just in time for the Seven Years' War, he produced the first of his American maps, entitled *A Map of the Most inhabited part of New England.* In about 1766, Jeffreys was joined by a young apprentice named William Faden, who, on Jeffreys's death, would lead the firm in becoming one of Britain's most prolific and respected cartographic businesses.

Another American map was produced by the Virginian John Mitchell (1711–1768), who had once studied medicine at Edinburgh. In his New World homeland, Mitchell not only practiced as a physician but also became an accomplished botanist and land surveyor. In 1746, when ill health forced him to seek medical care in London, Mitchell decided to exploit his knowledge of the colonies and pay for his medical bills by producing a map of his homeland. Entitled *The British and French Dominions in North America,* this map was probably the most reliable of the time and was used during the Anglo-American peace negotiations at the conclusion of the American Revolutionary War in 1783. Later, it was used again as a basis for the settlement of the boundaries separating the new United States of America from British Canada.

In the same year of 1759, fellows of the Royal Society of London, whose company was led by Bradley, were beginning to discuss an up-coming astronomical event of significant importance. In 1716 Edmund Halley, the second astronomer royal and Bradley's predecessor, had predicted that an opportunity would arise in 1761 to observe a transit of the planet Venus across the face of the sun. Occurring in pairs, separated by 8 years, these transits occur only once a century. A successful observation of the transit would allow scientists to determine the distance of the sun from the earth and provide scale to the size of the solar system. And while they were at it, the scientists would be able to determine the precise position of hundreds of seaports around the globe.

To be successful would require sending out expeditions to record, simultaneously, the phenomenon from various parts of the world. It would be a very expensive program of expeditions, and not all the academics agreed that it was worthwhile. Fortuitously, in the same year, Father Boscovich was visiting England as a guest of Bradley and other fellows of the Royal Society and added his considerable reputation behind those urging that the society to send observers overseas.

It may seem strange that the Royal Society and His Britannic Majesty's astronomer royal could contemplate sending expeditions abroad when Britain and much of Europe were at war. Indeed, many of Europe's overseas possessions and their indigenous populations were also in a state of violent flux. News from foreign lands traveled "exceeding slow" in those days. A letter to or from America could take several months; from the Orient, the better part of half a year or more. Thus it was never certain, in the more distant corners of the empire, who was the victor and who the vanquished. In part, the paradox is explained by the fact that in the eighteenth century war was a more gentlemanly thing; this is not to say that civilians were not routinely massacred or their goods and chattels plundered or destroyed. But, in a more general way, for the mass of the population, war passed them by.

It was not unusual for communications and even trade to continue between the populations of belligerent nations. The citizens of Dover, at the southeast extremity of Britain, maintained regular trade (much of it illicit) with their counterparts in Calais and openly ferried paying passengers across the channel during wartime. Letters continued to pass, after a fashion, between the fellows of the Royal Society in London and their colleagues in the Académie Royale des Sciences in Paris. In the early years of the disagreeable Napoleonic Wars, a similar degree of free communication was maintained, and the astronomer royal was able to ensure that his friends at the Paris Observatory received his important observations and copies of his *Nautical Almanac.*

There was also a protocol that scientific expeditions were exempt from attack during wartime. For example, when Alexandre-Gui Pingré set forth in November 1760 on his voyage to observe the transit of Venus from Rodriguez Island, the Académie Royale des Sciences applied to Britain for a laissez-passer. The letter, explaining his purpose and requesting any military unit "not to molest his person or Effects upon any account, but to suffer him to proceed without delay or Interruption,"[6] was readily granted. There are many examples of scientific instruments being seized by an enemy only to be passed on when their scientific purpose was discovered, "with his Excellency's compliments in expectations of your success." The pursuit of science tended to transcend the energy of war—but not always.

Where the Seven Years' War had taken William Roy away from his mapmaking in Scotland, it opened the door of opportunity to the naval officer Lieutenant James Cook (1728–1779). Whereas Roy was to pioneer the science of precise mapping on land, Cook was destined to do the same for charting at sea.

James Cook was born in the same year as Charles Mason and was just 18 months younger than William Roy. Unlike Roy or Mason, Cook came from the very humblest origins, the second son of James Cook and his wife, Grace. The elder James was a day laborer who had fled the penury of his Scottish homeland to settle in Yorkshire, where he found work with Thomas Scottowe. The benevolent Scottowe saw to it that the bright, young James Cook, Jr., received a rudimentary education. After a short time in the grocery trade, the young Cook

James Cook (1728–1779).

transferred his apprenticeship to a Quaker family by the name of Walker, who owned a fleet of colliers. Captain John Walker saw the boy's potential, took him under his wing, and taught him the basics of navigation.

In 1749, Cook left England for the Baltic trade, returning in 1752 as a ship's mate in Walker's fleet. He was offered command of his own ship in 1755 but, with the outbreak of the Seven Years' War, decided to join the Royal Navy as an able seaman. Cook was so skilled a seaman that within 2 years he had risen to the rank of master* of HMS *Pembroke*, a sixty-gun frigate.

In 1758, at the fall of Louisburg, the French stronghold commanding the approaches to the Saint Lawrence River in Canada, Cook's vessel was in nearby Kennington Cove. Surveying the cove that day was Lieutenant Samuel Holland, a Dutch émigré and military engineer who had served valiantly with the Royal American Regiment of Foot. Holland was using a plane table to map the cove, and Cook watched him through his spyglass. The sailor was so intrigued that he went ashore to observe the surveyor at work firsthand. The two men fell to talking, and Cook asked Holland if he would teach him land surveying. Holland was pleased to oblige and an appointment was arranged for the next day.[7]

Captain Simcoe, the *Pembroke's* scientifically minded commander, learning of the meeting, invited Cook to bring Holland and his surveying equipment

* Masters ran the sailing of a ship; the captain was in overall command.

aboard the ship. Holland had recently secured copies of captured French plans and documents relating to the Saint Lawrence and its vast gulf. Realizing the strategic advantage that reliable charts would give the British fleet as they prepared for the assault upon Quebec, Simcoe encouraged Holland and Cook to take the *Pembroke* upriver from Cape Breton and, using the captured charts, make new and more accurate versions.

The charts they produced were invaluable to the British fleet's commanders as they prepared for the assault on Quebec. Unfortunately, the charts did not cover the entire navigation of the river that the fleet would need. To solve this problem and to lead in the British ships, Cook and other masters of the squadron took soundings ahead of the ships along the stretch of the treacherous river known as the Traverses. Conducted entirely from open boats and under the guns of the French army, these brave men opened the way for Admiral Saunders's fleet to navigate right up to and past the enemy stronghold. This bold maneuver brought the British fleet close to the city and paved the way for General James Wolfe to storm the Plains of Abraham and secure the capture of the vital city.[8]

Following the fall of Quebec, Cook was transferred as master of HMS *Northumberland.* His new captain, Lord Colville, proved to be a very worthy patron and one who recognized Cook's outstanding abilities. Under Colville's encouragement and influence, Cook extended his charting adventures to include the rocky coasts of Nova Scotia and Newfoundland. When the *Northumberland* was dispatched to Saint John's, Newfoundland, in September 1762, Cook found himself working with another surveyor, with the grand name of Johan Frederick Wallet des Barres (1721–1824).

Like Holland, des Barres was an émigré, a Swiss mercenary who had volunteered for war service with the Royal American Regiment. He had learned his trade at the Royal Military College at Woolwich and in the employ of the Dutch army. Cook and des Barres made a good team; while the Swiss surveyed the abandoned French fortifications at Conception Bay with his plane table, Cook made charts of the anchorages, using his Hadley quadrant and des Barres's landmarks. Des Barres taught Cook the elements of trigonometry, and in return Cook passed on his knowledge of hydrography; both men learned a great deal of each other's trade.

After the fall of Quebec, des Barres remained in North America, where he was engaged by the British Admiralty to continue his surveys of the coasts of Nova Scotia and the harbors in Newfoundland. In 1763, he was appointed by the British Admiralty to expand his work to include the coasts of New England. In 1773, he returned to England, of which he was now a naturalized subject, to prepare and publish the results of his hydrographic labors. These were printed in

1784 as *The Atlantic Neptune,* one of the finest collections of sea charts of the century. In later life, des Barres was appointed lieutenant-governor of Cape Breton Province and, later, governor of Prince Edward Island. He died in Halifax, Nova Scotia, at the ripe old age of 103.

William Roy and James Cook were, essentially, self-taught surveyors, and each was to take his scientific arts to the highest levels of perfection. Both men enjoyed mathematics and shared a passion for physical astronomy. They were driven men, sharing many traits—natural leadership, unbounded energy, and dedication above and beyond their callings—and each was at ease with some of the finest scientific minds of the day. They knew and admired each other's work, and both were to become friendly with Nevil Maskelyne; in the case of Roy, the friendship was particularly strong.

The fall of the major French Canadian cities to the British land and naval forces marked the end of the French and Indian War (as it was known in America). In Europe, the Seven Years' War rumbled on. As the Royal Society's deadline approached for sending out its transit of Venus observing expeditions, there was still no sign of an end to the hostilities. In this dangerous climate of uncertainty, two expeditions prepared to set off. One was to go to Saint Helena, in the South Atlantic and was to be led by Nevil Maskelyne; the other was destined for the East Indies. The first expedition to hazard the seaways consisted of just two men: Bradley's observatory assistant, Charles Mason, and a land surveyor from County Durham, Jeremiah Dixon (1733–1779).

In late November 1760, Mason and Dixon left London by coach, with their chests of scientific equipment, bound for the south coast naval base at Portsmouth, Britain's premier naval port. Foul weather and other delays prevented their ship, HMS *Sea Horse,* from sailing until 8 January. When the storms had subsided sufficiently, Captain Smith navigated his sixth-rate frigate out into the choppy waters of the English Channel and set course to the west and the Bay of Biscay.

Their objective was Fort Marlborough, the East India Company's "factory" at Benkulen, in Sumatra, a backwater of the "honourable company's" great trading empire and a place with a dubious reputation. Unbeknown to anyone at the time, the status of Benkulen as a trading post and as a suitable place for peaceful astronomical observations had changed dramatically for the worse. In the face of a possible assault, its governor had "shamelessly" surrendered the entire settlement to a French naval flotilla.[9] In the event, it did not matter because far worse things were in store for the two intrepid astronomers.

Just 2 days after leaving Portsmouth, "at 11 o'clock in the morning," a French warship was seen bearing down on *Sea Horse* from windward. It was the French

*L'Grand,* a thirty-four-gun frigate, larger and more powerful than *Sea Horse.* As the two ships closed, a fierce fight broke out and the engagement raged for "1 hour 10 minutes"[10] before the Frenchman broke off and headed away. The *Sea Horse* set off in pursuit—"all the sail possible was made to keep up with her"—but the British vessel was badly mauled. Eleven of the ship's crew lay dead and many more were seriously wounded, "a great many of which are mortal," wrote Dixon. Smith's gallant little warship was forced to return to Plymouth for repairs and make a "press run" ashore to secure replacements for the depleted crew.

From the Devonshire naval yard, Mason and Dixon conducted an acrimonious correspondence with their employers, the Royal Society. The forced delay for repairs had made it impossible for the astronomers to reach the East Indies by June in time for the transit. Even had they done so, it is extremely doubtful the occupying French forces would have been sympathetic. Instead, Mason and Dixon argued the case for diverting to the Black Sea or to the Levant, which would have suited the requirements quite as well, but the Royal Society was adamant: it was either Sumatra or a court of law.

In the event, Mason and Dixon were shown to be correct. They were landed by Captain Smith at Cape Town, the Dutch colony in southern Africa where La Caillé had studied the southern constellations just 3 years before. The two British astronomers observed the transit successfully, remaining for several months to observe and record many new "navigation stars" for Bradley's star catalogue. In the meantime, their colleague Nevil Maskelyne was ensconced on the East India Company's Atlantic way station of Saint Helena.

Maskelyne's transit of Venus had been a disaster. The almost perpetual cloud over the rocky island had obviated any chance of observing Venus. He and his assistant, Richard Waddington, did, however, conduct experiments to see if they could measure the theoretical attraction of mountains. That Maskelyne failed in the attempt was not because there was no effect but because his instrument, a 12-foot zenith sector built by the celebrated instrument maker Jonathan Sisson, suffered from an alignment problem. Maskelyne was not a man easily disheartened by a faulty instrument and instead turned his energy and considerable intellect to other physical measurements. As soon as the *Mercury,* an East India Company "snow" (a small, fast packet boat), delivered Mason and Dixon to Saint Helena on 16 October, the two astronomers were immediately conscripted by Maskelyne to assist with gravity experiments.

Thanks to the French discovery, 80 years before, that gravity had an effect on the oscillating rate of clocks' pendulums, the method had become widely accepted as a means of determining small gravity differences. Mason and Maskelyne began their observations, using a special clock called an astronomical regu-

Cape Town. Illustration from Charles Darwin's
*Journal of a Voyage Round the World,* 1835.

lator, made by the London instrument maker John Ellicott. Dixon returned to
Cape Town with another clock, made by the renowned clockmaker John Shelton, to take simultaneous readings "that we might have a double proof of the
action of Gravity in different parts of the Globe on the length of Pendulum."[11]
(Shelton's clock we shall come across again.) To fill the hours between gravity
observations, Mason and Maskelyne measured the levels of the Atlantic tide
and drank their way through the incredible quantity of liquor that the Cambridge don had deemed necessary to take with him.

Several months later, their work done, Mason and Dixon left Maskelyne and
Waddington on Saint Helena and set off for England aboard the *Prince Edward.*
By the time they arrived in London on 6 April 1762, they and their exploits were
well known. At about this time, Thomas Penn and Lord Baltimore (represented
by his uncle, Cecilius Calvert), two very rich and powerful men, were engaged
in resolving a long-standing and ofttimes violent dispute over the boundaries
separating their American provinces of Pennsylvania and Maryland. At stake
were some 4,000 square miles of territory and thousands of pounds in un-
collected taxes. Lord Harwicke, sitting in the court of chancery, had ruled on
the dispute some years before, and the disputants had accepted his resolution to
survey and settle their boundaries. In accordance with the ruling, commissioners had been appointed by both parties to oversee the necessary work.

Despite the best efforts of local American surveyors, establishing boundaries according to the court's instructions had proven to be somewhat difficult. The survey task was monumental and extremely complex. In the event, the governors of the two provinces decided to write to their proprietors, asking for modern surveying tools, while at the same time appealing for professional assistance. At the time, there was no one person in England with sufficient skills and knowledge of surveying and astronomy to solve the problem of the boundaries. But just then along came Charles Mason and Jeremiah Dixon.

# 9

## VERY EXPERT IN HIS BUSINESS

THE LONGED-FOR PEACE WITH FRANCE was signed in February 1763, and the hostilities that were the Seven Years' War came to an end. That is, the hostilities between the European powers ended, but in America the disaffected natives who had supported the French, led by their great chief Pontiac, continued to wage a campaign of violence against British forts and European settlers all along Pennsylvania's western marches. It was in this atmosphere of fear and retribution that, on 15 November 1763, Mason and Dixon arrived in Philadelphia.

The story of Mason and Dixon's extraordinary work and exciting exploits is recorded elsewhere,[1] it is enough to say that for the next 5 years these two men, Protestant astronomer and Quaker land surveyor, worked in perfect harmony, surveying and setting out the longest and most precise boundary lines ever drawn. Between them, they developed new methods that combined astronomy with land surveying—methods that would be emulated and improved by others in the years to come. In addition to their unprecedented boundary surveys, they also, for the Royal Society, were the first British scientists to measure a meridian arc; and they did it in America. They also spent a freezing Pennsylvanian winter observing the first gravity measurements taken in North America, and for this purpose Maskelyne sent them an astronomical regulator—none other than John Shelton's clock, which they had used on Saint Helena and at Cape Town.

The land they chose for measuring the meridian arc was the lowland of what is now Delaware. The swamps, groves of tall cedar, and slave plantations stretched across a landscape as "level as a floor by nature." The entire distance from the Delaware middle point marker to their headquarters at John Harlan's farm in Chester County had already been measured once, when setting out the borders separating Maryland from the "three lower counties" (Delaware). Triangulation was pointless and, in any case, would have been almost impossible across the densely wooded and flat landscape. Instead they carefully remeasured their original lines and diligently reobserved the latitudes.

The 100 miles were measured by special measuring rods made by Joel Baily, one of their surveyors, and carefully calibrated steel "Gunter" survey chains. No one had ever before attempted to measure such a long and accurate line, and the principles they and Nevil Maskelyne (who had charge of the project in London) developed would be of direct importance to William Roy and other earth measurers in the future.

Before leaving London, the physician and keen amateur astronomer John Bevis (1693–1771) calculated from Bouguer's 1749 figure of the earth that at the latitude where Mason and Dixon were working, the equivalent ground distance of 1 degree of latitude should be 68.72 miles (110.6 km). However, when Mason compared the distance they had actually measured across the flat landscape with the latitudes that he and Dixon had observed with John Bird's 6-foot zenith sector at the terminals of the baseline, he discovered that 1 degree of latitude was actually 68.86 miles (110.8 km).

Mason was not surprised to discover the difference, some 725 feet (221 m) greater than predicted: "By comparing our mensuration of a Degree of the meridian with that made under the Arctic Circle [by Maupertuis], supposing the Earth to be a Spheroid of a uniform Density. But the Earth is not known to be exactly a Spheroid, nor whether it is everywhere of equal Density."[2] In this brief statement, written in a Pennsylvanian farmhouse in 1768, Mason succinctly summed up the contemporary state of geodetic knowledge. Even at this early date, Charles Mason understood the problem well enough, perhaps one of only a handful of scientists who did. Variations in the earth's density were affecting the direction of the zenith sector's plumb line, and it was the insidious "attraction of mountains" that was to blame.

However, back home in England, Nevil Maskelyne, who succeeded as astronomer royal on the death of Dr. Nathaniel Bliss, Bradley's short-lived successor, did not concur. The nearest mountains were the Alleghenies, and they were 100 miles west of the meridian arc. In his mind the discrepancy of 725 feet had to have another cause "as there is no room for suspicion that the plumb-line of the sector could be deflected materially from its proper position by the attraction of any mountain."[3] It took a more gifted scientist to come up with a very reasonable supposition. Maskelyne's curious and incredibly shy friend and colleague, Henry Cavendish (1731–1810), pointed out that not only might mountains affect the direction of the vertical but also a "defect" of attraction could be caused by the depths of the Atlantic Ocean to produce a similar but contrary result. Cavendish estimated that the two contesting phenomena combined might lead to an error of between 380 and 636 feet (116–194 m) over the distance

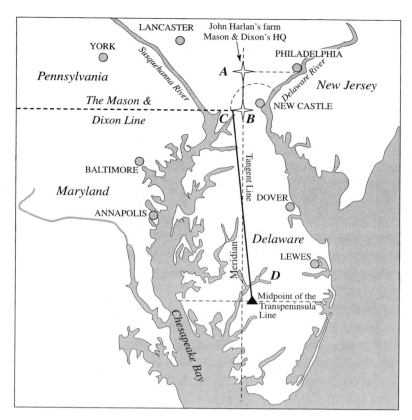

Mason and Dixon's meridian arc in America.

of 69 miles. The phenomenon now became "the attraction of mountains and the deficiency of oceans."

In the same year that Mason and Dixon went to America, "it came under the consideration of Government to make a map of the whole island [Great Britain] from actual survey."[4] The idea of the military men of the Board of Ordnance was to incorporate the Watson-Roy prewar map of Scotland within a complete map of the British Isles, and William Roy was to direct the operations.

Roy had been steadily improving his mapping skills and surveying knowledge, especially astronomy and triangulation, and was already friendly with Nevil Maskelyne and the scientists of the Royal Society. In July 1765 he was appointed "Surveyor-General of Coasts and Engineer for making and directing Military Surveys in Great Britain."[5] The appointment, by Royal Warrant, did not go so far as to give Roy the authority to execute the entire survey of Great Britain, but from this time onward, he used every available opportunity to pursue his mapping exploits in preparation for the great things that were to come. For these duties, the master-general of the ordnance paid Roy an additional 20 shillings (£1) a day; a considerable sum that continued until Roy's death.

Roy and the many influential supporters of the mapping project, including the astronomer royal and the Royal Society, were persistent in their lobbying of government to fund properly the entire project. As in past conflicts, the Seven Years' War had again shown the value of good maps and sea charts for military maneuvers and logistics—and had also demonstrated the consequences of not having them. The military was desperate for a complete map of Britain, but the government was broke, or at least too badly strapped for cash to fund a full-scale project. They were not unaware that Cassini's great map of France, begun in 1736, had had to be rescued from fiscal embarrassment by private enterprise and, consequently, had become a readily available product for any army, friendly or hostile, to acquire.

Perhaps the British government did not see the benefit of funding a project that private enterprise was already fulfilling, albeit slowly, as a result of the 1759 initiative by the Royal Society of Arts. Whatever the reasons, and despite the support of the king, "the consideration of Government" was not to divert its scarce funds for a military map of the islands; in any event, within 7 years of its "consideration," the start of hostilities with America would put an end to the scheme. Nevertheless, Roy's reputation as a scientific mapmaker was widely acclaimed, and in 1767 his endeavors were rewarded with a fellowship of the Royal Society.

Mapmaking apart, 1768 was another year of fevered scientific preparation. The second transit of Venus was predicted to occur in the following June, and scientists throughout Europe and in America were making their preparations. The 1761 event had not been without its problems, but many practical lessons had been learned and there was an international determination that the next, and final, transit for 100 years should be a spectacular international success. All told, some 150 teams from France, Germany, the Scandinavian countries, Russia, Italy, and across the rest of Europe would be mobilized to observe the event. In America, the members of the newly founded American Philosophical Society of Philadelphia were preparing for their grand debut on the stage of international science.

In Britain, preparations were growing apace. James Short (1710–1769), the Surrey Street instrument maker, astronomer, and fellow of the Royal Society, was given the task of compiling all the observations. British parties were to be dispatched to Hudson's Bay, Ireland, Norway, Lapland, and the Pacific Ocean, and instruments were to be sent to William Hirst in India for making the observations there. Father Boscovich, then professor of mathematics at the University of Pavia, was invited to lead the British expedition to Spanish California. Unfortunately, Boscovich felt he was unable to accept because of the great unpopularity of the Society of Jesus, to which he belonged, and other political constraints.

At the council meeting of the Royal Society on 14 January 1768, Nevil Maskelyne read a paper discussing the instruments that would be needed for observing the transit and for determining latitude and longitude at the various widely separated locations. Sitting on the council that evening was the great American scientist and astute political lobbyist Benjamin Franklin. His presence in London was, officially, to represent his home province of Pennsylvania in its struggle against the Penn family's monopoly of power and to petition King George III (who succeeded his father in 1760) for a change in the province's status to a "royal colony."

This was Franklin's second official residency in Britain, having previously left, with some reluctance, in 1762, when he became frustrated by the government's stubbornness in refusing to understand the colonists' viewpoint. Despite their disagreements with the way Britain administered its American interests through the proprietorial system, the colonists were, by and large, loyal to their king. Indeed, many in England were equally happy: "You know that we have a new King and a new Parliament," wrote Samuel Johnson to his Italian correspondent Joseph Baretti in Milan. "We were so weary of our old King, that we are much pleased with his successor; of whom we are so much inclined to hope great things, that most of us begin already to believe them. The young man is hitherto blameless; but it would be unreasonable to expect much from the immaturity of juvenile years, and the ignorance of princely education."[6] Johnson's gentle way with the new king was flavored, perhaps, by the fact that George III had just bestowed on him a pension of £300 a year. But, for a time, Johnson would be satisfied with his new monarch.

Benjamin Franklin had lodgings with a widow, Mrs. Stevenson, and her daughter, Polly, at 36 Craven Street, off London's Strand. During his brief return to Philadelphia, he had remained in correspondence with them both. He loved England and the English and, had it not been for a tragic turn of events, would have settled down in London. Writing from America to Polly Stevenson, he stated, "Of all the enviable things England has, I envy it most its people. Why should that petty island, which compar'd to America is but like a stepping stone in a brook, scarce enough of it above water to keep one's shoes dry; why, I say, should that little island enjoy in almost every neighbourhood more sensible, virtuous and elegant minds, than we can collect in ranging 100 leagues of our vast forests."[7] Franklin's energy and reputation were indispensable in securing for the Royal Society a grant of £4,000 for the transit campaign from King George.

In early 1768, James Cook had also returned to London, fresh from his Canadian hydrographic charting surveys. Two years before, he had observed a solar eclipse from Burgeo Island (the same one observed by Mason and Dixon) and had derived a precise longitude for Newfoundland. His observations had been sent to his acquaintance Dr. John Bevis, who communicated the results to the Royal Society. Bevis was well acquainted with Cook's expertise, referring him to his colleagues as "a good mathematician and very expert in his business."[8] The British Admiralty also had good reason to be pleased with Cook for the work he had done in opening the way for the invasion of Quebec. Cook appeared to be the ideal man to make the transit observations in the Pacific. However, he had a rival, Alexander Dalrymple, who thought his credentials more worthy.

Like Roy, Alexander Dalrymple (1737–1808) was a Lowland Scot, born at Newhailes (East Lothian). Unlike Roy, Dalrymple came from the Scottish aristocracy; his brother was David Dalrymple, Lord Hailes (1726–1792). In 1752, Dalrymple decided to seek his fortune in the service of the East India Company and in 1753 secured a position as a "writer," or clerk, in Madras. It was while working in the company's offices that the young Dalrymple became interested in the heaps of maritime journals that lay around the offices of the "honorable company." From these journals and from his astute observations, he deduced that the company was not doing well in China; in fact, it was not doing any business in China at all.

Dalrymple thought he detected a solution to the problem among the hundreds of sea captains' journals. His brilliant idea was to set up a trading "factory" not in China itself but at a location where the enterprising Chinese could come and go and where Chinese national sensibilities could be preserved. He brought his ideas to the attention of the company's council, who, somewhat desperate for some additional income, set Dalrymple off to see what he could find.

In 1759, Dalrymple took the schooner *Cuddalore* into the Sulu Sea and explored the regions north of Luzon before making his way back to Macao, charting the coasts as he went. In 1761 he was back in the Sulu Sea, surveying and trading his way down to the Philippines. Dalrymple's exploits and the mass of knowledge he collected on these voyages were beginning to show results, and his employers were very pleased with him. In 1762, as Charles Mason and Jeremiah Dixon were plowing northward from the East India Company's way station of Saint Helena, Dalrymple again set off in the *Culladore* to negotiate with the sultan of Sulu and set up a trading factory. On this occasion, the explorer took along with him a 20-year-old naval midshipman, James Rennell, the future and first surveyor-general in India, of whom more later.

The following year, Dalrymple returned to England for what he thought would be a short break; but despite living to a ripe old age, he was never to return to his beloved Indies. Instead, to offset his disappointment, he began to put his knowledge and records to use in the East India Company's offices in Leadenhall Street.

On the subject of exploration, Dalrymple was perhaps one of the widest-read individuals in Britain. He was a firm believer in the existence of a "great southern continent," as were many other distinguished persons, and his extensive charting activities in the oceans and seas of the Far East made him a very knowledgeable scientist. The Royal Society recognized in Dalrymple the natural candidate to command the transit of Venus expedition to the Pacific. Unfortunately for Dalrymple, the Admiralty was not so enamored of the man, despite his achievements. Their Lordships were mindful of a previously unfortunate experience, when the astronomer Edmund Halley had commanded one of His Majesty's ships, and they were adamant that they would not tolerate ever again "any philosophers" commanding another. Admiral Hawke went so far as to say that "he would rather cut off his right hand than permit anyone but a King's Officer to command one of the ships of His Majesty's Navy."[9] In fact, the Admiralty had unjustly vilified Halley, but in the situation over the appointment of Dalrymple for the Pacific, it was probably correct. It would prove to be an adventure that would tax the capabilities of the most skilled seaman.

In the end the Admiralty had its way, and on 4 May 1768 the 40-year-old Lieutenant James Cook was given the command of the joint Admiralty-Royal Society Transit expedition to the Pacific. Charles Green (1735–1771), Charles Mason's friend and replacement at the Royal Observatory, was to be the astronomer. However, this was not the end of Alexander Dalrymple's part in the story of the earth measurers, and we shall return to him later.

The day after Cook's appointment, Captain Samuel Wallis returned to England after a long voyage of discovery in the Pacific. Leaving his vessel, the *Dolphin,* in the downs off the naval port of Deal, he rushed to London, bringing exciting news. He had discovered a new island in the Pacific that he had named King George's Island (Tahiti). Wallis also reported that he had seen, far away on the horizon, the unmistakable telltale signs of the fabulous "Great Southern Continent." Wallis's report changed everything. The plans for a casual voyage of discovery were frantically revised and Cook received new top-secret, sealed

The *Endeavour* in harbor for repairs, an engraving by W. Byrne after Parkinson.

orders, to be opened only after he had completed the transit of Venus. The orders instructed him, in the king's name, to conduct a thorough exploration of the southern oceans, to find the continent, and to claim it for Great Britain.

The first priority remained—Cook had to get to the Pacific in time to observe the transit. From Wallis's enthusiastic description of his new island, King George's seemed the perfect spot. Cook's old commander and friend Admiral Palliser had purchased a ship of the type he knew Cook would approve, a Yorkshire collier named *Earl of Pembroke*.[10] Renamed *Endeavour*, Cook's cat-built Whitby bark would become one of the most famous of all ships of exploration. However, its fitting out in the Deptford yards of Mr. Bird was nearly a catastrophe.

The rising cost of food after the war, a succession of poor harvests, and the stagnant and even declining wages were matters for serious unrest throughout the land. In London, the weavers of Spitalfields rebelled, and the sailors and coal-heavers working London's River Thames emulated the weavers' example. "Riots and tumults" erupted up and down the river, the men's ill temper exacerbated by the unusually hot weather. Shipping came to a halt and prices soared. Sailors demasted and derigged their ships while others deserted. The protests, strikes, and outbreaks of civil unrest that swept across England during the hot summer of 1768 were the zephyrlike harbingers of Engel's "veritable storm."

For three weeks, the *Endeavour* languished in the heat on its slipway and "received much damage from it."[11] The seams in the planking opened so wide in places that a man could put his fist between the boards. While the *Endeavour* sweltered and shrank in the London heat, the frozen north was the destination for the first transit observers to depart for foreign shores. The two men selected by the Royal Society to go to Fort Churchill on Hudson's Bay were the astronomers William Wales (1734–1798) and Joseph Dymond.

Hudson's Bay was frozen solid for much of the year, so in order to be ready for observing the transit in June 1769, which occurred before the ice was clear for sailing, they had to leave a year early. The prospect of overwintering in northern Canada was not to Wales's liking; he had requested a "warm climate," hoping to be appointed to the Pacific expedition. Dymond and Wales joined the Hudson's Bay Company's trading ship in Great Yarmouth on the bleak shores of East Anglia on 23 June.

A month later, the *Endeavour*'s problems were over and it was free to leave the slipway at Deptford and move out into the Thames for fitting out. The price of the vessel and the fitting out had cost the Admiralty £8,194, 15s, 4d—a considerable sum for the time. Hundreds of people visited the ship, and sumptuous dinners were held in the great cabin for the well-to-do and "persons of quality." Samuel Johnson was one of the visitors, relating afterward that, had he been to sail on the ship, he would have "turned from her with relief to gaol."

Others were not as critical as Johnson, and among the "supernumeraries" that Cook would take with him on the great voyage of discovery was the amateur botanist Joseph (later Sir Joseph) Banks, "a gentleman of large fortune," and his eight "companions."[12] There were also the distinguished Swedish botanist and natural philosopher Joseph Solander and, of course, astronomer Charles Green and his servant James Reynolds. Captain Cook, together with the "gentlemen" (the officers) and the supernumeraries occupied the stern half of the ship, while the *Endeavour*'s crew of eighty-five sailors, "the People," as Cook called them, were crammed into the fo'csle and along the gun deck.

The ship sailed from the Thames without Cook and anchored in the downs off the naval town of Deal to take on stores. Cook and Green traveled to the little town from London by coach and were rowed out to join their ship. With Lieutenant James Cook in command, the *Endeavour* set course for Plymouth, where it was to collect the young and, according to some accounts, "pretty" Joseph Banks and his large party. Finally, on 26 September, the crew of the *Endeavour* bid farewell to England and set sail into the Atlantic, bound for the distant coasts of South America. Nearly half of the men would never see home again.

# 10

## A PASSAGE WITH MY HORSE

AS *ENDEAVOUR*'S BLUFF BOWS BATTERED through the Atlantic swells south-ward toward Cape Horn, the Falmouth packet bearing Mason and Dixon left the busy waterfront of New York for its 5-week passage to England. Mason and Dixon were back in London, lodging at the Prince of Wales Arms in Leicester Fields (Leicester Square), in time for Christmas. They had a busy work schedule, writing papers, assembling their notes, reporting to the proprietors, and set-tling the accounts. On 10 November, they presented a paper to Nevil Maskelyne recording their scientific endeavors for the Royal Society.

Ten days before Christmas, Mason and Dixon attended the regular meeting at Crane Court to present their paper to the society's fellows and their guests and to hear the latest plans for the transit of Venus expeditions. The turnout that evening was most satisfactory, watched by Ben Franklin and a nervous Charles Mason; the "noble lords" arrived in their coaches, escorted by liveried footmen, drawing into the narrow confines of Crane Court. Some gentlemen ar-rived by "chair," the famous if somewhat leaky sedan chairs that cluttered Lon-don's thoroughfares, while the rest of the audience ambled in from the local inns and hostelries of Fleet Street.

Accounts of Mason and Dixon's exploits and scientific achievements in America were widely discussed and the surveying methods they had pioneered meticulously examined. Both men had been invited to join the transit of Venus expeditions and now just awaited the details. At the meeting, Dixon learned he was to be teamed with William Bayly (1737–1810), Charles Green's replacement as "labourer" at the Royal Observatory. They were to be the observers for the North Cape of Norway, the very apex of Europe. Bayly was yet another example of brilliance rescued from rural obscurity. The son of John and Elizabeth Bailey (the name spelling varies), tenant farmers of Bishop's Cannings in Wiltshire, William began life as a humble plowboy. He was befriended by a local excise man (an unpopular official in the eighteenth century), who recognized the boy's talents and gave him rudimentary lessons in arithmetic. Bayly's aptitude with numbers came to the attention of William Kingston of Bath, who secured him a position as an usher at a school in Stoke, on the outskirts of maritime Bristol, and soon young Bayly's mathematical prowess became widely known. By 1768 he was enjoying the patronage of the powerful duke of Richmond, a senior figure on the Boards of Ordnance and Longitude, and in the same year was engaged by Nevil Maskelyne as his assistant at the Royal Observatory. With additional funding from the duke in return for mathematical assistance, Bayly's basic salary of £26 soared to £60. A week after Dixon and Bayly's appointment, Maskelyne addressed the council and proposed Charles Mason as the astronomer for the Irish observations.[1] However, over a month was to pass before Mason was formally appointed. In the meanwhile, he returned to Gloucestershire to join his family at the house of his old friend Robert Williams[2] and his daughter Mary.

On 1 March Mason once again packed his traveling chests and set off from London for Ireland by way of Ludlow, Chester, and the Dee-side port of Parkgate. Embarking on Captain Lloyd's packet, *Venus,* they set sail across the Irish Sea to Dublin. From the capital of Ireland, Mason made his way on horseback, first to Londonderry, where he presented his credentials to the authorities, then on to County Donegal, choosing Cavan, near Strabane, as a location for his temporary observatory.[3]

A month later, on 13 April, Jeremiah Dixon and William Bayly sailed with Captain Douglas aboard HMS *Emerald,* bound for northern Norway beyond the Arctic Circle. Bayly was dropped off first, at the North Cape, on 28 April, together with his equipment and temporary housing. Dixon went ashore on 7 May at Hammerfest Bay, where the sailors helped him erect his prefabricated "house" and observatory amid the desolate, snow-covered rocks.[4]

The next party to leave was John Bradley, a nephew of Samuel Johnson, and Captain Hunt, the master of HMS *Arrogant,* who sailed from Portsmouth on 11 May aboard the sloop *Seaforth,* bound for Falmouth, Cornwall.

Preparations also went ahead at Greenwich and at many of the private observatories around Britain. Every nation in Europe was preparing, and those with overseas missions or foreign interests sent out teams of scientists. But in British America, political relations with the home country were getting worse by the day. As Charles Mason boarded the ferry to Ireland, the merchants of Philadelphia, some of them members of the American Philosophical Society, joined in the widening and popular boycott of British trade goods, luxuries, and slaves. A month before the date of the transit, George Washington read out

George Mason's "resolutions" to the Virginia House of Burgesses, raising the cry of "no taxation without representation!" In England, Samuel Johnson responded with his own Tory philosophy of "Taxation no Tyranny," countering that taxation is the right and duty of any government to raise money to protect the freedom and liberty of its people. In his defense of the measures, he remarked wryly, "How is it that we hear the loudest yelps for liberty among the drivers of Negroes?"[5]

As in times of war, science transcended civil disobedience and trade boycotts. The arguments over the rights and wrongs of taxation were matters of politics and not of science. It is indeed wonderful to look back on how men found it so easy to differentiate among commerce, politics, and "interests" in both their conduct and their correspondence. Benjamin Franklin's letters to his friends in London during the troubles leading up to and during the American Revolution succinctly demonstrate this delightful and whimsical gift.

In the worsening political situation, the American Philosophical Society determined to remind the home government that American science was as good as anything in Britain. A Transit Committee was formed, and a special observatory was built in the yard of the State House in Philadelphia. Instruments were purchased, loaned, and borrowed. Thomas Penn, rising above the petty squabbles with his legislators, graciously provided the society with his zenith sector and the transit and equal altitude instrument that Mason and Dixon had used on the famous boundary survey. He also bequeathed to them a new brass telescope made especially for the event by James Short. Penn also acted as an intermediary among the American Philosophical Society, the Royal Society, and the Astronomer Royal in London to coordinate efforts and share information.[6]

The transit was also an opportunity for the American Philosophical Society to add an accurate longitude to Mason and Dixon's latitude of its bustling city. In addition to the new observatory at the State House, two other observing sites were set up, partly to compensate for bad weather and partly to measure longitudes. The first of these was at David Rittenhouse's plantations at Norriton, and the other was at Lewes in Sussex County, near Cape Henlopen. David Rittenhouse (1732–1796), provost of the University of Pennsylvania, a distinguished mathematician, astronomer, and instrument maker, was put in charge of the work.

When the great day of 3 June 1769 arrived, hundreds of observers scattered about the globe were ready to observe the transit of Venus and to determine the latitudes and longitudes of their locations. Most were lucky with the weather, but some were to be disappointed. For the first time in history, a concerted international effort added a hundred or so precisely and scientifically located points around the world to which the old maps could be corrected and on which new ones could be founded. Many discoveries were made, including that of David Rittenhouse, who claimed the discovery of an atmosphere about the silver planet.

On King George's Island, or Otaheite, as Cook more correctly called what is now Tahiti, a wooden palisade enclosing a small makeshift observatory was constructed on a promontory at the north end of the island. Here at Point Venus, Cook, Charles Green, Joseph Banks, and Dr. Joseph Solander set up their quadrants and telescopes. One of the Royal Society's astronomical regulators was firmly embedded into the sand to aid with the time measures and to determine the strength of gravity. When eventually Nevil Maskelyne received copies of the records, they showed that the clock was losing at a rate of 1 minute, 20.9 seconds of mean time per day. He was able to remark, "The same clock, when fixed at the Royal Observatory at Greenwich, before the voyage, with the pendulum of the same length, got at the rate of lm 45.8s per day, on mean time, between April 19 and July 18, 1768. Therefore the force of gravity at Greenwich is to that at King George's Island, as 1000000 to 997075." The inference was simple: on Tahiti, the strength of gravity was less than in London, and hence Tahiti was further from the earth's center. Whatever the reason for La Caillé's mysterious discovery at the Cape of Good Hope, it could now with some certainty be dismissed as an aberration: the Cook party had substantiated the Newton-Bouguer oblate spheroid hypothesis *south of the equator*.

On Tahiti, the latitude was fixed from "meridian zenith distances of the sun and fixed stars" and found to be 17°29'15.4" south. In reviewing the latitude observations, Maskelyne was moved to reflect, somewhat uncharitably, "It must be confessed, that the results of these observations (most of which were made by Mr. Green) differ more from one another than they ought to do, or than those do made by other observers, with quadrants of the same size, and made by the same artist, the cause of which, if not owing to want of care and address in the observer, I don't know how to assign."[7] The longitude for the island was measured from lunar distances and "eclipses of Jupiter's satellites." It is interesting to note how accurate these measures could be: the longitude by the former "navigator's" method was 149°36'38" west of Greenwich and by Jupiter's moons, 149°32'30", a matter of just 4.5 miles.

The transit of Venus over, one by one, the overseas contingents began to arrive home laden with astronomical information and gravity observations ready for James Short to start working. On 30 July, Dixon and Bayly arrived at Sheerness, in the Thames estuary. Two and one-half months later William Wales and Joseph

Dymond sailed into Plymouth harbor from freezing Hudson's Bay, just in time for an exceptionally cold British winter. By and large, the European effort to observe the transit had been a success. Some, however, had mixed fortunes.

The French astronomer Jean-Baptiste Chappe d'Auteroche had one of the longest and most difficult journeys, which took him to Spanish Baja California via the bandit-ridden interior of Mexico.[8] On reaching his objective, the Spanish mission at San Jose del Cabo, he discovered that the villagers were dying from an "epidemical distemper," in short, typhus. Despite the tragedy happening all around him and discounting his own fears, he managed to observe and fastidiously record the entire transit. Within days, his traveling companions fell to the killer disease, and he bravely tended them as best he could until he too succumbed and died. His records survived and were eventually sent to Paris, where an examination revealed that Chappe's work was probably the most precise of all the observations made in 1769.

Charles Mason stayed on in Ireland until 28 November, taking many astronomical observations, including an appearance of Halley's comet that heralded the birth of two baby boys who would one day face one another across the battle-scarred fields of Waterloo. Mason's return journey was more dramatic than the outward trip. From "Donahadee [Donaghadee, Co Down]" on New Years' Day, he "attempted to make a passage with my Horse to Port Patrick in Scotland [Rinns of Galloway], in a small open Boat, but when we were about 1/3 over, a Gale of Wind arising with a high Sea, we were obliged to return and with difficulty reach'd our port."[9]

Why Mason felt it expedient to return from northern Ireland (with a horse in an open boat) by way of Scotland and in the middle of winter is extraordinary unless there was some serious point to such a dangerous trip. He was certainly determined; the next day he fared better: "Went aboard a large House Boat, and had a pleasant passage,—reach'd Port Patrick at ½ past 6$^h$ PM being about 4½ on the passage." Mason's notes of his return trip south from the Scottish lowlands hold some clues: "From Port Patrick in my way to London past [sic] through Dumfries, Carlisle, Penrith Kendal Lancaster Preston Warrington Newcastle under Line Litchfield Meriden Coventry and reached London Saturday the 13th day of January 1770 at 6h PM."[10] The long, bitterly cold ride through the hills and glens of Galloway and down the west coast of England took him through the snowy mountains of Cumbria and within sight of the Derbyshire Peak District. The harsh journey was by way of a scientific reconnaissance. There can be little doubt that he was in fact preparing himself for a very important experiment, and although the diversion was worthwhile and relevant, it had serious consequences for Mason's already delicate health.

Meanwhile, the excitement of the transit over, all that remained for the scientists was to complete the calculations and present the papers. Charles Mason, recovered from his arduous horse ride, presented his report to the Royal Society in person on 25 January 1770. In the council minutes for 1 February, Nevil Maskelyne praised Mason's excellent work and the quality of his observations and for his "masterly report." Dixon and Bayly's paper was also included among the many reported during that busy month. The first results from America had also arrived and were read out to the packed meetings by Benjamin Franklin.

In the same month, Lord Frederick North, the eldest son of the earl of Guilford, a staunch Tory, succeeded the duke of Grafton as prime minister. Seriously worried by the growing unrest in the American colonies, he repealed the unpopular Townshend Duties but left untouched the unpopular tax on tea. In believing that this move would go some way in appeasing American anger, he was badly mistaken. Although a competent Tory politician, witty, good-humored, and a close friend of King George, North's vacillating administration (1770–1782) would lead Britain into war.

In early March, Maskelyne learned to his satisfaction that the distinguished gentlemen of the American Philosophical Society in Philadelphia had, on 18 January, added his name to that of Mason and Dixon by electing their astronomer royal to membership. Father Boscovich also visited London and joined in the discussions over the results from the transit of Venus. On one notable occasion, he met Samuel Johnson at Mrs. Cholmondeley's house, a favorite haunt of the literati. Neither speaking the other's language well, they conducted their conversation entirely in Latin. They discussed the "superiority" of Newton and the recent war with France. On the latter point, Boscovich laughed aloud when Johnson jokingly suggested that "we had drubbed those fellows into a proper reverence for us, and that their national petulance required periodical chastisement."[11] Both men were in fact very fond of France. They met on a number of occasions at Sir Joshua Reynolds's entertaining dinners and at the bishop of Salisbury's London mansion. Their correspondence continued even after Father Boscovich left his home in Italy to take up King Louis XV's offer of the post of directeur d'optique au service de la marine, a position that came with the handsome salary of 8,000 francs. In 1783, Boscovich, in ill health, returned to Milan, where he died in 1787.

The advantage offered by the fevered astronomical climate was taken by the astronomer royal as an opportunity to determine the positions of the observatories in British America with respect to Greenwich. On 2 August 1770, Nevil Maskelyne wrote to his new colleagues in America, care of Thomas Penn, requesting that their observatories should be linked together by a framework of survey lines and, in turn, to the Royal Observatory at Greenwich by simultaneous longitude observations. This suggestion was quickly agreed to as colonial mapmaking was also on the minds of the American Philosophical Society, and they set about the task with great energy.

In Lewes (Delaware), Mason and Dixon's right-hand man, Joel Baily, together with Owen Biddle and Richard Thomas were busy surveying. "We employed ourselves," wrote Biddle to the astronomer royal, "in measuring the distance of our place of observation from the stone fixed at the beginning, or east end, of the east and west line, which is the boundary between the three lower counties and Maryland, and is situate on Fenwick's Island; the latitude and longitude of this place [the Delaware middle point marker] being accurately determined by Messieurs Dixon and Mason."[12] Mason and Dixon had not in fact located the stone on Fenwick Island, but the astronomer royal was able to reply that he had spoken with Jeremiah Dixon, who had provided the distance from the Fenwick stone to the middle point, "sixty nine miles wanting 100 yards," which was sufficient.

To measure the longitude, the appearance and disappearance of Jupiter's moons were observed at Greenwich, at the State House observatory in Philadelphia, and at John Rittenhouse's plantation at Norriton, Pennsylvania. Thus, by default, the longitude (and time) of British America was tied irrevocably to that of Greenwich. Eventually Greenwich was to become the origin for longitude all over the world, but at the time, the French counted their longitudes from Paris, the Spanish from the Canaries, and so on.

Tied by politics and proprietorship to Britain, America's links with the mother country were further tightened with this, the first scientific transcontinental geographical connection. The State House observatory in Philadelphia, built by subscription especially for the transit of Venus, had served its purpose except for one last, and perhaps its most important, purpose—from its wooden stage on 4 July 1777, the Declaration of Independence was first read out to the people of America the day after its secret signing. In the event, longitude was the only tie that was to endure.

About the time of Boscovich's visit to London, Jeremiah Dixon decided to retire from active science. It would seem that family matters and the coal business needed him back in County Durham. Mason returned to Gloucestershire and reunited with his two young sons at the Williams's household in Tetbury. It would be nice to believe that Dixon accompanied him on one last expedition, when Charles Mason married Mary Williams in the pretty Queen Anne church of Saint Kenhelm, Sapperton.

The transit of Venus of 1769 marked a pivotal point in the science of earth

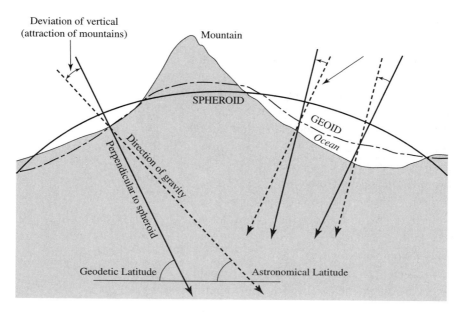

The attraction and "deficit" of mountains.

measurement. The shape of the earth was now definitely agreed to be oblate, at least north of the equator, thanks to the work of the French and Italian academicians and, more recently, that of Mason and Dixon in America. In the light of the Cook expedition's results on Tahiti, most scientists regarded La Caillé's arc in South Africa as an anomaly for further investigation. Not only was the shape of the earth settled, more or less, but more important its principal dimensions were now known.

One remaining piece of information was still missing, which scientists were determined to resolve. Evidence from the many gravity observations and the small inconsistencies in some of the meridian arc measurements strongly hinted that the density of the earth, still unknown, was not completely consistent with a solid or even hollow sphere. These inconsistencies, it was believed, were synonymous with the deviant effect of the attraction of mountains and were blamed for corrupting the accuracy of latitude measurements.

Newton's predictions about the shape of the earth had been proved to be correct, as the effect of some of his gravity theories demonstrated. The great man's prediction about the attractive force of mountains, as well as Henry Cavendish's calculations suggesting a reverse effect caused by the ocean deeps, could not be ignored and marked the genesis for a new theory about the nature of the earth's crust that we now know as isostacy. Pierre Bouguer's attempt to detect the effect on Chimborazo in 1736, although inconclusive, had established a viable method for experimenting. All that was missing was a very accurate astronomical instrument with which to measure it.

The quest to measure the attraction of mountains was no idle scientific wild goose chase but an experiment of the most profound importance. If the density of the earth, whatever it turned out to be, showed regional variations from what a homogenous oblate spheroid should exhibit, then any measure of latitude, and longitude, come to that, would contain some small unknown error. A deviation of just 10 seconds of arc in latitude, the least, some argued, that the effect would induce, would result in an error of over 1,000 feet (300 m) on the ground. Even the positions of the observatories that determined the positions of the stars in the night sky might be in error. All maps and charts that relied on precise latitude and longitude would also be in error and, worse, by amounts that varied from place to place.

Bouguer's *La Figure de la Terre* had been published in Paris in 1749, since when more measurements had been made. From studying its pages, Astronomer Royal Nevil Maskelyne concluded that Newton's hypothesis for the effect of mountains on the plumb line[13] was probably the cause for the inconsistencies that the French scientists had seen in their arc measures. Maskelyne had attempted to detect the effect for himself in 1761 among the mountains of Saint Helena. On that occasion, it had been the plummet alignment of his zenith sector that had let him down. As a consequence of this experience, he and the London instrument maker John Bird (1709–1776) had designed an improved plummet arrangement. Mason and Dixon's experience in America with a 6-foot Bird sector fitted with the new plummet had demonstrated the extraordinary accuracy these giant, unwieldy instruments could achieve in skilled hands; and whose hands were more skilled than those of Mr. Charles Mason? It was at this point that Maskelyne decided that, if anyone was going to measure the attraction of mountains, it was going to be a Briton and it would be Charles Mason supervised and guided by His Majesty's astronomer royal.

On 2 May 1770, Maskelyne proposed to the council that the Fleet Street instrument maker, John Bird, should be engaged to fit the society's own zenith sector with the new plummet arrangement. This was the instrument, made by the late Jonathan Sisson, that Maskelyne had used on Saint Helena. The council concurred with the request, as he knew it would. With the measuring means within his grasp and a candidate to do the work, Maskelyne went away to consider his proposed experiment in detail and, in particular, the nagging question of how to do it and where on earth he was going to find a suitable mountain.

# 11

## FRANKENSTEIN AND OTHER EXPERIMENTS

FOR THE REVEREND NEVIL MASKELYNE, 1771 was another busy year. James Cook and his ship, the *Endeavour,* returned to Britain from the Pacific in July, and the success of the expedition to the Pacific spurred the idea of yet another voyage of discovery, despite the tragic consequences of the first voyage and the death of the Greenwich astronomer Charles Green. In fact, half the crew and "gentlemen" had died, not from scurvy, but from malaria contracted in Batavia (modern Jakarta). The new voyage would involve two ships: Cook in the *Resolution* and Tobias Furneaux in command of the *Adventure*. Once again, astronomers were to be included, and once again Nevil Maskelyne found himself much engaged in the preparatory work.

In December, Maskelyne proposed to the Royal Society that his astronomical almanac computer, William Wales (the late Charles Green's brother-in-law), who had observed the transit of Venus from chilly Hudson's Bay, and the astronomer William Bayly, who had accompanied Dixon to Norway and was Green's replacement at the Royal Observatory, should join the two ships as astronomers and instructors in advanced navigation techniques. Bayly's replacement at the observatory was a truculent Yorkshireman named Reuben Burrow, of whom we shall hear more. It was also agreed at the meeting that the second Cook expedition, as it was beginning to be called, much to the annoyance of Joseph Banks, should be equipped with the very latest technology for finding the longitude.

Copies of John Harrison's celebrated invention, the chronometer, were ordered to be made; these were extremely expensive and exquisite examples of the horologist's art, over £60 each, and to some extent their use would make or break the new-fangled longitude solution the "mechaniks" had dreamt up. Neither Cook nor the Admiralty would be disappointed with the results from "the marine clocks."

William Roy and Nevil Maskelyne were, if not already friends, well known to each other by this date. Roy, who is probably best remembered for his work on the Roman military antiquities of Scotland rather than as a master surveyor, was in Edinburgh that October 1771. He was conducting a trigonometrical survey—"for amusement," as he idly called these campaigns—to locate the position of a new observatory at Hawkhill, on the northeast side of the Scottish capital, for Maskelyne's friend James Lind, a man who would have an essential but largely forgotten role to play in future events.

The actual purpose for the survey is unclear—perhaps they were trying to determine a relative position against some known point—but what is certain is that both Lind and Roy shared the task of surveying. On 26 October, Roy wrote to Lind explaining that his, Roy's, calculations for the survey could contain a mistake as he had been in a hurry. The triangle they had observed together, the observatory, the great rocky promontory east of the city known as Arthur's Seat, and Calton Hill, contained, said Roy, an error of 5 minutes of arc (1/12 degree) over the required 180 degrees. Roy laid the blame for the error on his telescopic quadrant.[1] The exact site of the Hawkhill observatory is not certain; the city's first official observatory was designed by James Craig and was erected on Calton Hill in the 1790s, superceded in 1894 by the Scottish Royal Observatory.

There were at the time in Edinburgh two James Linds; Maskelyne's friend should not be confused with his more famous younger cousin, the physician who discovered the cure for scurvy that James Cook did not use. Maskelyne's James Lind, M.D. (1736–1812), was an amazingly skinny individual, "a mere lath," as his father in-law, Dr. Charles Burney, the musicologist and friend of Samuel Johnson, described him. He was in every respect a most delightful man: witty, possessed of a "sweet disposition," and a respected medical practitioner who would one day become physician to the royal household at Windsor.

Madame d'Arblay (1752–1840), Burney's novelist daughter, laughingly called Lind "a better conjurer than a physician."[2] Magic was not his only hobby for Lind was also a renowned astronomer and had observed the transit of Venus from Edinburgh. Despite Lind's apparent homely nature (he was an avid antiques collector, a passion shared by Roy), he was another example of that strange and delightful conundrum the eighteenth century was so good at creating. By the age of 26, he was in Bengal, practicing, so it seems, as a "surgeon." In those happy days, a surgeon was quite a long way down the ladder from a qualified doctor of medicine. Nevertheless, he was already accomplished at his art and even assisted during the fever epidemic in Bengal of 1762.

Four years later, Lind was working as a surgeon on an East Indiaman, where he most likely first became enamored of astronomy, learning something of the science from the ship's navigators in between sawing the limbs off damaged sailors. He returned to his native Edinburgh to qualify as a physician in 1768. In comparison with the willowy doctor, Madame d'Arblay described Anne Elizabeth Mealy as Lind's "fat handsome wife, who is as tall as himself, and about six times as big."

In his later days in Windsor, after his wife's death, Lind was a mentor to some of the boys of nearby Eton College, among whom was a young, aspiring poet by the name of Percy Bysshe Shelley. The schoolboy was much impressed and influenced by the good doctor, and Lind makes an appearance as the old hermit in Shelly's *Laon and Cythna* and as Zonoras in *Prince Athenase*. Lind's interests were extremely eclectic and delightfully eccentric; apart from astronomy and surveying, he designed the first rifled cannon and was an early pioneer in forensic medicine. He was also a close friend and confidant of that most celebrated mechanical engineer James Watt, whose steam engines were to revolutionize, indeed drive, the Industrial Revolution of the latter half of the century.

In 1789, Lind participated in the exhumation of the body of King Edward IV at Windsor Castle. There, Lind had examined the remains and analyzed the lethal liquids trapped within the lead-lined coffin. He also experimented with "galvanism," wiring up dead frogs to electrical Layden jars or to Franklin's sparking electrical machine and observing the reptiles' jerking muscles. All this fascinated the young Shelley, and later in life he would reminiscence about Lind's gruesome pastimes. It was on one of these occasions that the 19-year-old Mary Woolstoncraft, staying with her lover at Byron's house on Lake Geneva, was inspired through a nightmare to unite Lind's grisly researches with literature. The result was her famous gothic novel, *Frankenstein* (1818).[3]

It may have been through Maskelyne that William Roy first became acquainted with James Lind, or maybe it was that Roy knew him in some other connection, such were the ways of things in that curious century. Men who indulged in "the sciences" were few and far between, and despite the vast gulfs that often divided them socially, they seemed to have formed a sort of brotherhood of mad scientists. When it came to discussing science or swapping notes, it mattered not if the writer was a factor's son, like Roy; from the squirearchy, like Maskelyne; a baker's boy; the bright offspring of a desperately poor Scots laborer; or an immensely rich landowning peer of the realm, like the president of the Royal Society. The delight of new learning was a great leveler. Of course, the social distinctions remained inviolate, but the important point—and just one of those little idiosyncrasies that kept the English aristocracy from enjoying the terminally unpleasant consequences suffered by their French cousins—was that there was an almost, but not quite, profound appreciation for intellectual equality.

The other great leveler of the time was cricket. Gentlemen of quality played the game with their tenants, and as in science, the rules were simple: the best player (independent of rank) was the captain, and his word, on the day and on the pitch only, was law: "Sorry your Grace, but you're out!" Not only did these quirky games and scientific interests occasion intercourse, however fleeting, between the classes, they also were the continuation, almost the offspring, of that precious commodity called freedom that Englishmen were not averse to expressing, if necessarily by force.

In the latter half of eighteenth-century England, unlike much of Europe, freedom meant not having to fight for your lord; indeed, it meant being able to exist *without* dependence on some great man. It meant being able to wander at will wherever and whenever you wished (as long as you had money in your pocket); it meant being able to take a job and, when fed up with it or a better offer came along, to quit. Freedom did not mean you had the vote; that was simply a matter of having sufficient wealth. Freedom allowed the opportunities to exist that made acquiring wealth open to all men, even of the humblest origin, and also to women as long as they were not too desperately humble or trying to bring up a family of sickly children on sixpence a week.

Freedom in the Age of Reason also allowed men to express their feelings and political views more openly and without danger of punishment (at least by the law). Gentlemen traded insults and ridiculed one another through articles in the press (Benjamin Franklin being one of the most proficient at this sort of freewheeling libel) or in anonymously published pamphlets, the authors identifying themselves as Greek sages or Roman poets; their true identity was blatantly obvious to everyone.

The press and pamphleteers were not just amusing sideswipers; they were also very serious and, for political thought and scientific progress, irreplaceable prognosticators and kindlers of the most vital sort. Pamphlets also offered opportunities for friends to trade insults or question the veracity of one another's ideas in a semianonymous way without creating a row across the dinner table. For example, when Nevil Maskelyne lambasted the deceased Charles Green for his scruffy notekeeping on Cook's first voyage, Maskelyne's "esteemed friend," James Cook, immediately responded:

> Mr M should have considered before he took upon him to censure these observations, that he had put into his hands the very original book in which they were written in pencil, only, the very moment they were taken and I appeal to Mr M himself, if it is not highly probable that some of them might from various causes, as either to be wholly rejected

or to be marked as dubious and which might have been done had Mr Green taken the trouble to enter them in the proper book. Mr M should also have considered, that this was, perhaps, the only true original paper of the kind ever put into his hands; does Mr M publish to the world all the observations he makes good and bad or did he never make a bad observation in his life."[4]

By the time William Roy was pointing his quadrant toward Lind's observatory at Hawkhill, his ambition to perfect his surveying knowledge had already extended to the higher mathematics and triangulation. Like the French of an earlier generation, he had come to appreciate the value of the method for creating the framework upon which to hang his mapping campaigns. Roy and Lind had much in common in the way of mapmaking. At some time, possibly the previous year, Lind had fixed the latitude of Islay, off the Scottish west coast, and acquired a "beautiful map of the island."

Roy's appointment as surveyor general of coasts in 1765 had been expanded to include the duties of deputy quarter-master general. All this kept him very busy, so he was seldom to be found in his seat at the Royal Society. In 1769, he had been instructed to draw up a report on the defenses of Britain's Mediterranean outpost at Gibraltar and make recommendations for their improvement. It is a moot point, but it seems unlikely that he ever actually visited the Rock; rather he made use of the many drawings, sketches, and maps that then existed.

Ten days after James Cook and his ships sailed away on their second great voyage of discovery on 13 July 1772, Roy was present when Nevil Maskelyne presented his proposal for the gravity experiment to the Royal Society. Impressed and excited by the idea, the council established a formal committee of members "for considering some experiments which are proposed to be made for measuring the Attraction of Hills and Mountains."[5] There can be little doubt that Maskelyne had already consulted Roy and Charles Mason on a suitable location for the experiment. After all, Mason's arduous trek through Scotland during the winter of 1769 was no idle journey.

In 1770, Charles Mason took up piecework for the Board of Longitude, and supervising the computations for the *Nautical Almanac* through Maskelyne, who was a member of the board. On 7 February 1772, Maskelyne wrote to Mason "c/o Robert Williams, Tetbury," anxious about work falling behind on his precious almanac: "I must have some calculations from Mr Mapson [one of the almanac computers] very soon." Mapson was a friend of Mason and lodged with their old headmaster, Robert Williams. While Mason was away in America, Maskelyne had written to Mr John Nourse, "Bookseller, Near the New Church, In the Strand. Sir, I desire you will send a nautical almanac stitched in blue, directed to Mr. John Mapson at Mr. William's Tetbury Gloucestershire, to the Old Change, to go by Mr. Creed's Waggon." With typical Maskelyne attention to minutiae, he continued, "The Waggon sets out Thursday or Friday Morning, I suppose early. . . . Do it up in stiff brown paper folded two or three times, or it would be better to put it between pieces of board."[6]

In March 1772, Mary Mason had given birth to a son, John, and, on 13 March she and her husband proudly took him to be baptized at St. Kenhelm. Four months later, James Lind set sail with Joseph Banks on a voyage of botanical adventure to Iceland. In Gloucestershire, Mason's life settled down to a happy humdrum existence, and very soon young Mary was pregnant again. The next year, Mason's American friend Dr. John Ewing arrived in England, and there was the occasion for a happy reunion, some good dinners, and a chance to catch up with the latest gossip from the colonies. That all was not well between Britain and its American interests was clear to anyone who could read a newspaper.

The European population of the American colonies had already passed 2.5 million, and the political demands for recognition were growing louder. Samuel Johnson, an avid Tory, wrote of the American population explosion, "Their numbers are, at present, not quite sufficient for the greatness which, in some form of government or other, is to rival the ancient monarchies; but by Dr. Franklin's rule of progression, they will, in a century and a quarter, be more than equal to the inhabitants of Europe. When the Whigs of America are thus multiplied, let the Princes of the earth tremble in their palaces."[7] Even more chillingly, Johnson prophesized, "If they should continue to double and to double, their own hemisphere would not contain them. But let not our boldest oppugners of authority look forward with delight to this futurity of Whiggism."

With dissent escalating among the settlers, especially in the big towns, there was a danger that things could get out of control. In January 1770, 6 months after the transit of Venus, New York's Sons of Liberty had clashed with forty British soldiers, and there were serious injuries on both sides. This skirmish paled into insignificance when word reached London of the so-called Boston Massacre in March, when five men taunting British redcoats were shot dead. The colonists were only slightly mollified when the soldier's commander, Captain Thomas Preston, was arrested along with eight of his men and charged with the murders. At last, it appeared that the British government was beginning to understand its American subjects.

At the trial of Captain Preston, the American lawyers John Adams (destined to become the second president of the United States) and Josiah Quincy, both active in the independence movement, successfully defended the British captain and six of his soldiers on the murder charge. Two men were found guilty of manslaughter and were branded before being freed.

In Britain, Parliament passed the Regulating Act for India with the intention of controlling the more political activities of the East India Company and to solve its desperate financial difficulties. The Tea Act of the same year was the government's attempt to improve the "honourable company's" tea sales, principally to America, whose people were refusing to buy British tea because of the high tax. The result, the famous Boston Tea Party, led to Prime Minister North's response: enactment of the despised Coercive Acts.

All these events were followed closely by Benjamin Franklin, still languishing in London as the representative of the Pennsylvanian Assembly but essentially enjoying himself. He was present on 24 June 1773 when the Royal Society met to hear Nevil Maskelyne's progress report and consider his latest ideas for the great gravity experiment. The portly astronomer royal explained to the fellows how he needed to find a hill, a very special hill, so that the experiment could have the best chance of success. He had been making various inquiries, to Lind in Edinburgh, who had only recently returned from Bank's expedition to Iceland, and through Thomas Hornsby, his astronomical colleague at Oxford's Radcliffe Observatory. He also, of course, had Mason's report from his 1770 winter journey. And there was Franklin's account of his journey through the Pennines with the chemist Ingenhousz and John Canton in 1771, where they had visited the high peaks and the deep limestone caverns while "fossilizing," or searching for minerals.

On Maskelyne's recommendation, having primed the fellows on a number of previous occasions, the committee "ordered that Charles Mason be employed in the Observations respecting the attraction of Mountains in Scotland."[8] At this point, the council of the society expected Charles Mason to conduct the experiment himself and, it can be concluded, the aging astronomer had already agreed to it with Maskelyne and William Roy. The astronomer royal's papers and the records of the survey themselves betray many things that reveal the hand and astute mind of Mason at work. All went well for Mason's and Maskelyne's plans, and on 18 July the society passed a resolution to send, and pay, Mason to embark on yet another journey of discovery. His orders were to travel to the remotest parts of the country—to Yorkshire, Lancashire, and the Highlands of Scotland—to find a suitable mountain "of sufficient height tolerably well detached from other hills and considerably larger from East to West than from North to South."[9]

# 12

## A REMARKABLE HILL

ON 3 AUGUST 1773, NEVIL MASKELYNE wrote from his study at Greenwich "To Dr James Lind, Edinburgh, North Britain" with good news.[1] Their Oxford colleague, Thomas Horsley, had recently received an account from "his correspondent at Perth," James Robinson, who had made a survey of Glen Tilt. The glen had been recommended by William Roy and was an early contender for the gravity experiment. From the soldier's description of Glen Tilt, Maskelyne was hopeful that the gravitation effects of the great defile would be sufficient to deflect the plumb line by about 16 arc-seconds, equivalent to some 1,600 feet (488 m) on the ground, well within the accuracy capabilities of the zenith sector that would be used for the experiment.

However, reading the report of Horsley's correspondent at Perth, Maskelyne wryly noted that the dimensions of the glen had shrunk by "2/3 of what Col Roy had estimated them."[2] Whether Roy himself had actually inspected the glen or had resorted to his "magnificent military sketch" is unclear. In his letter to Lind, Maskelyne went on to defend his estimable friend, suggesting the error was the result of just being an estimate. But an error of such magnitude was quite incredible for a surveyor of William Roy's distinction.

Glen Tilt was no longer a suitable contender for the gravity experiment, but it didn't matter because, as Maskelyne explained to Lind, Charles Mason had "this day set off by direction of the Royal Society," heading for Scotland and carrying "some portable instruments" to find a suitable alternative. The instruments Mason strapped to the back of his packhorse included two "barometers of Mr Luc's new construction," a theodolite, and, probably, the astronomers' workhorse, a Hadley quadrant. Mason's instructions, wrote Maskelyne, were to go first to Edinburgh to meet Lind and then to proceed to "Fort William & Ben Nevis &c to survey some of the principal hills in the Highlands."[3] Mason was further instructed, he informed Lind, to "determine the level of the country," using the barometer all along his route from Edinburgh to wherever he found the "tolerably detached hill." Without doubt, this request reflects the keen interest of Roy, who was a pioneer of measuring heights with a barometer.

Such a raft of information would be of no practical use to the experiment but would be of significant interest to Roy. Maskelyne provided Lind with detailed instructions on how to calibrate the barometers, the inference being that Lind would keep one of the delicate instruments at his observatory or house to use as a "base" to which Mason's heights could be corrected. The barometers measured atmospheric pressure; as this changed from hour to hour, by recording the changing pressure at Lind's house, the pressure measured by the "roving" instrument could be later adjusted.

In his letter, the great astronomer went on to discuss some of the practicalities; it is clear that he and Lind had been discussing Lind's involvement in the experiment for some time. Lind was requested to consider any possible hills that "he or his acquaintances might know" and to suggest some routes that Mason might explore once he arrived in the Scottish capital. Lind was requested to prepare a letter of introduction for Mason that he could use to secure accommodation and assistance from "some persons in his route."

The experiment, Maskelyne explained, required finding an oblong hill or a long valley oriented east-west or thereabouts and 1/2 mile high or deep or nearly so," about 2 or 3 miles long, but "no more than 5 miles." Despite the disappointment of Glen Tilt, Maskelyne still preferred a valley because then the two observing stations would be intervisible. If only a hill could fit the bill, then, wrote Maskelyne, it should be reasonably isolated from its neighbors, especially from those to the north and south.

The astronomer royal listed meticulously the deviations he expected the attraction of mountains to produce. Resorting to Newton's famous theory, he predicted "that a hemispherical hill, 1/2 mile high and 1 mile in diameter" would "offer: At the base, 12.4", at 1/2 mile distant, 3" and at 1 1/2 miles distant, 3/4 "."[4] Mason's own suggestion for measuring the variation in gravity of a mountain was "by the going of a clock," that is, observing the effect of the mountain's mass on the rate that an astronomical regulator's pendulum would oscillate, a method that Bouguer had employed in the Andes and which would be used again in 1854 by a future astronomer royal. Mason put a lot of faith in the mechanical excellence of clocks, especially the Shelton unit he had used in America, but his idea was not pursued.

Mason's instructions from the astronomer royal stated that he should make Ben Nevis his first objective. On Tuesday, 3 August 1773, Charles Mason set off "on horse-back with his instruments" up the Great North Road toward Scotland and was expected to arrive in Edinburgh about 11 August. Also heading north that month was an expedition led by Captain the Honourable Constantine Phipps of the *Racehorse,* together with Captain Skeffington Lutwidge of the *Carcass.* Their objective was to sail "towards the North Pole"[5] to undertake scientific observations and charting. Mason's colleague Israel Lyons, one of his *Nautical Almanac* computers, was the expedition's astronomer, and among the "young gentlemen" (as cadet naval officers were called) was the 14-year-old midshipman Horatio Nelson.

Mason had traversed the great Atlantic four times; he had visited exotic Africa, roamed the untamed wildernesses of America, tussled with Indians, been shot at by the French, and enjoyed the remote hospitality of County Donegal. Now he was to return north and penetrate the "most uncivilised Parts of the Highlands of Scotland." What Mason's thoughts were as he set off with his packhorse and scientific payload bound for the fastness of "the wild Irish" we may never know. Perhaps he felt a little trepidation; at 46 he was already getting on in age, and the prospects of 8 or 10 days in the saddle could not have been attractive. His long journey north took him past the home of his old friend Jeremiah Dixon in County Durham; perhaps he took the opportunity of dropping in for a few bottles of claret and a good long chat. Within 5 years, Dixon would be dead.

When setting off on such a journey as Mason's, a common practice of the day was to buy a horse at the beginning and then sell it at the end. For example, when Dr. Skene of Aberdeen traveled from London to Edinburgh in 1753, he found that "being nineteen days on the road, the whole expenses of the journey amounting to only four guineas. The mare on which he rode, cost him eight guineas in London, and he sold her for the same price on his arrival in Edinburgh."[6] Mason could not be certain of the treatment he might encounter along the way. Often as not, country yokels were hostile and rude to any stranger, as William Hutton from Birmingham discovered to his cost in 1770: "The inhabitants set their dogs at us in the street, merely because we were strangers. Human figures not their own are seldom seen in these inhospitable regions [Warwickshire!].[7] Surrounded with impassable roads, no intercourse with man to humanise the mind, nor commerce to smooth their rugged manners, they continue the boors of Nature."

A few months before Mason set off, a smart post chaise had transported Samuel Johnson to Edinburgh to meet his young Scottish friend, James Boswell. The objective of these two Highland tourists was to visit the Hebrides by way of Aberdeen and the coast road around to the Great Glen (at the western end of which stood Ben Nevis), then to follow Bonnie Prince Charlie's escape route to Skye. Like Mason, Boswell and Johnson needed both guides and interpreters. In

Inverness, Johnson hired riding horses for himself, Boswell and their servant, plus "two Highlanders to run beside us, partly to shew us the way"[8] and to act as interpreters.

Arriving in Edinburgh, Mason found Lind's home among the elegant houses of the newly built Prince's Street, in the new town. The old town, whose fortress General Wade had, 50 years earlier, secured from the advances of the city's ladies of the night, lay on the far side of the deep defile that now houses Edinburgh's Waverly railway station. The remaining details of Mason's sojourn in Scotland and his exploration of the hills of Yorkshire have to be filled in from contemporary observations as no papers have been discovered. From Maskelyne's letter, we know that Lind had been asked if he would be "so kind to instruct him [Mason] how he will best go about the language of the Highlands." The "language of the Highlands" where Mason was going, was Erse, the Gaelic tongue brought to the land by Kenneth mac Alpin and the Scotti (Irish) invaders of the eighth century.

Mason spent several days with the skinny Lind and his fat, jolly wife, going over his exploring plans, reviewing the latest list of likely hills, and discussing the logistics of the expedition. No doubt other Edinburgh worthies showed interest in Mason's reconnaissance and the scheme he and Maskelyne had hatched for discovering the elusive attraction of mountains and the density of the earth. If Mason was on form, there would have been some entertaining dinners, copious bottles of wine, and much scientific discourse interposed with tales of foreign climes, sea battles, slave traders, and savage Indians.

By the middle of August, Mason was on his way toward the mysterious fastnesses of the Highlands. From the intelligence gathered from the "philosophers of Edinburgh," Mason's primary object was clear, and it was not to be Scotland's most noble peak. Maybe it had been Hornsby's correspondent of Perth, Lind, or one of the Edinburgh experts who suggested that Mason would be likely to find a "suitable hill" for the experiment among the mountain peaks beyond Perth. Instead of heading for Ben Nevis, as instructed, he directed his horse along the southern shores of the Firth of Forth, crossing by way of Queen's Ferry to the north shore and a short ride north to the city of Perth.

"The people go about barefoot," wrote Johnson, "to shew that shoes may be spared." What the great lexicographer omitted to mention was that poor English children did the same, and for similar reasons. Yet, in the impoverished Highlands, even the children of the relatively well-to-do went barefoot. But if Mason expected to see signs of "black meall," he would have been disappointed: oatmeal was beyond the pocket of the poor, whose main nutritional supplement was, as Johnson noted, "kail" (a type of cabbage).

At Perth, the gateway to the Highlands proper, Mason visited Thomas Hornsby's "correspondent," James Robinson, to pay his compliments and to get some firsthand knowledge of the hills to the north. Robinson probably assisted Mason in securing the services of a guide to look after the astronomer as he wandered among the hills and glens and to act as an interpreter with the natives. On the road northward into the hills, they passed small fields of corn tucked between rough stone walls and tiny patches of potatoes, "but lately introduced," standing beside the simple cotts the locals called home. The people, observed Johnson of the Highlanders, "want [lack] the conveniences of common life," but he was pleased to note, things had greatly improved since 1745. Instead of finding that "their tables were as the feasts of Eskimeaux, and their houses filthy as the cottages of Hottentots,"[9] Mason discovered a somewhat better state of affairs.

Poverty was still extreme, but there was much less of it than he had probably expected. The women dressed openly in plaids and "affected the Highland manners" in defiance of the strictures of the, by then, redundant Proscription Act. "Civility seems part of the national character of Highlanders," observed Johnson, just as it does today among the folk of the mountains: "Every chieftain is a monarch, and politeness, the natural product of royal government, is diffused from the laird through the whole clan." Johnson was a Tory, as Mason probably was, and it had been the Whigs who had exacted such harshness on the Scots. Johnson's trip around the Highlands modified the bumptious doctor's earlier preconceptions that Scotland was a barbarous waste. His cruel but witty observation, when informed that Scotland had "a great many noble wild prospects," had been, "I believe, Sir, you have a great many. Norway, too, has noble wild prospects; and Lapland is remarkable for prodigious noble wild prospects. But, Sir, let me tell you, the noblest prospect which a Scotchman ever sees, is the high road that leads him to England!"[10] It was still true that a Scot's best chances for furtherance lay south, following one of Wade's roads into England, or west across the wide Atlantic, but things were improving.

Two or 3 days after leaving Lind, with the Royal Society's barometer hanging in the hallway of Lind's elegant Edinburgh home, Mason arrived in Aberfeldy, the last town of any consequence on his journey into the unknown. A few hours ride north lay his craggy objective. The first sight he had of lofty Schiehallion was in descending General Wade's military road from Glengoulandie, a view "that strikes the imagination with all the gloom and grandeur of Siberian solitude." Where Mason stayed is not recorded; perhaps he made use of his letter of introduction from Lind and stayed with the laird of Castle Menzies, or perhaps he took a room at the travelers lodge known as the White House, just above Kynnarchan. The opening up of the Highlands by Wade and Caulfeild's network of roads had greatly facilitated the "expeditious marching from place to place" of the Highland people, and the hills were not short of wayside inns. Even General Wade's old hut, used as his temporary headquarters during the road building, was "now a house of entertainment for passengers . . . not ill stocked with provisions."

Early next morning, Mason set out with his guide to reconnoiter Schiehallion. They followed the fast-flowing Tempar Burn southward, climbing up to the saddleback, where the 3,500-foot-high mountain (1,060 m) cast its deep shadow across the land. Looking northward, 2 miles distant, Mason could see the River Tummel, a silver ribbon of water running east from Loch Rannoch to Loch Tummel. The ground around the mountain was rough grass strewn with tumbled, moss-covered rocks and coarse heather. Turning southeast, they skirted the precipitous sides of the peak. Below them, the juvenile Allt Mór tumbled its rushing waters over rocks and stones in its narrow bed. Emerging from the morning shadow, the two men entered Gleann Mór, a broad space 200 or 300 yards wide (ca. 250 m), hemmed between the rugged sides of Schiehallion on the left and rounder, heather-clad hills on the right.

The flatter land of the glen was pitted with boggy pools and soggy peat beds, where Mason's horse sank to its fetlocks in the spongy mire. Following the Allt Mór, they passed by waterfalls and the mysterious Tarnnuarrea, the stone ruins of the ancient pastoral houses known as "schielings." They were now at a point beneath the great ridgeback where Mason thought he saw a likely spot for one of the observatories that would need to be built for the experiment. It was a narrow ledge, about 300 or 400 feet (ca. 100 m) above the valley floor.

Dismounting, Mason took measurements with his quadrant and made sketches of the land, his riding boots sinking into the boggy soil and filling with water. Despite the warm sun, a chill breeze ran through the glen. Hunting birds cruised across the steep flanks of the lonely mountain. The champing of the horse on his bridle and the lonely cry of the pipits were the only sounds to invade the silence of the secluded glen.

Smaller hillocks and outcrops encircled the base of the lofty peak and offered vantage points from where the astronomer could study the mountain and gauge its general form. Climbing up the rocky slopes of Mulichnahaoidenmore (Aodann Mór), Mason was able to get a better view of the ledge he had spotted earlier and of the waters bubbling down from the magical Well of Schehallien. To the east was another natural vantage point, Firroch (Fireach), which provided a double vista—westward along the glen and north toward a saddleback ridge where Schiehallion's slightly gentler eastern slopes met those of the rocky

Schiehallion. © M. J. Smyth, Institute for Astronomy, University of Edinburgh.

Tomnael, an outcrop of bare stone beneath craggy Dun Coillich. It was late evening by the time Mason completed his first circuit of the mountain. Leaving its painful beauty wrapped in deep purple shadow, he returned to his lodgings in time for supper, a pipe of tobacco, and a glass of whisky before tumbling wearily into bed.

The following day was even more arduous. Mason and his "ghillie" rode to the northeastern quarter of the mountain, where a path of sorts led up the long bare slopes toward the rocky crags that surmounted the "Fairy Hill of the Caledonians," as Schiehallion translates from the Erse. The way to the top was a narrow ridge, a rocky and treacherous route made even more difficult encumbered as they were with the theodolite and the delicate barometers. His Scottish companion shared the task of carrying the instruments. Mason was a man used to difficult terrain and hard-living people: he would not have shared Johnson's opinion that the Highlanders were "not commonly dexterous: their narrowness of life confines them to a few operations, and they are accustomed to endure little wants more than to remove them."[11]

Arriving at the first eastern peak, they stopped to rest. The view was breathtaking and brought back memories of the panoramas Mason had seen in the New World when he and Dixon had looked west from the peaks of the Allegheny Mountains across the forests of Pennsylvania to the distant plains of Ohio. Very few men could boast of seeing that sight. Perhaps, as they rested on the windy mountain, eating biscuits and cheese, Mason related some of his adventures to the ghillie—of Indians stalking the dark forests, the charms of Philadelphia, or the evening sun glinting on the distant Potomac.

The last stage of the climb brought them to the bare summit of Schiehallion, a desert of broken stones and sharp, rocky outcrops, blasted by the unfettered winds and shattered by the frosts of winter. All about them the mountains of the Highlands, ridge upon ridge, retreated in majestic splendor until they vanished in a horizon of blue haze. It was an extraordinarily beautiful sight; but there was work to do. The theodolite was unpacked and the business of surveying the peak began. It could only be a rough sketch at best, but it would serve its purpose.

De Luc's barometer was set up with the greatest care to avoid breaking the delicate glass tubes. The Scots ghillie watched bemused, scratching his mop of red hair, while Mason waited for the instrument to settle before recording the height of the mercury. Like the Indians of the American forests, Mason's companion found the scientific instruments very curious, maybe even slightly ridiculous, especially the three-legged stand that supported the brass circle of the theodolite that his strange English master kept stopping to peer through at goodness knows what. Only Mason's spyglass would he have recognized for what it was.

It required a week or more to explore the mountain and take the necessary measurements before Mason was satisfied that Schiehallion was the peak of choice. During that time, the mountain demonstrated its capricious character. Whereas one day would be sunny with racing clouds in a blue sky, the next could be heavy rain and dark gray clouds that enveloped the peak, hiding it from view. At such times, he was marooned; any work on the mountain was impossible.

Finally, the day arrived when Mason was satisfied he had enough information. He completed his calculations, collected his drawings, and made a sketch map from his copious notes. The last job to be done before he could leave the unpredictable, frequently miserable Highland climate was to set up a mark on the side of the mountain where, in his opinion, the first observatory should go. This done, and with a last backward glance at the great peak, Charles Mason mounted his horse and headed back toward Perth.

What Mason did next is speculation. From Maskelyne's papers to the Royal Society, it would appear that his instructions to visit the hills of Yorkshire and Lancashire were followed, although whether he bothered to retrace his steps and traipse all the way north to Ben Nevis is most unlikely. There is the impression that our indomitable hero was less than enamored with what he had found in Scotland. He was, after all, getting on in years, and the rigors of the hills and the Highland climate were not the kindest to a man whose constitution had already been tried to the limits.

Meandering down the west side of England, following the same road as years before, Mason took a look at Whernside, Wharfdale, and "Pennygent" (Pen-y-Ghent) in the Peak District to ensure he'd found the best option in Schiehallion. No doubt he was relieved to discover that, as candidates for the gravity experiment, every one of England's glorious peaks was a hopeless contender. From the Peak District, Mason followed the road south, passing through the town of Manchester, where he would have seen the grimy sights of the Industrial Revolution in all their grim detail. By 1774, Manchester was already a manufacturing metropolis of some 40,000 people engaged in the woolens industry, tinplating, and the new cotton business. Powered by coal, Manchester had already earned its reputation as the "smoky ward," but, more vitally, the bustling town and its environs were at the forefront of industrialization. It was the eighteenth-century equivalent of Silicon Valley. Mason would have seen, crossed, and wondered at the newly completed Bridgewater Canal, the world's first modern artificial waterway, completed in 1761 by the ingenious but illiterate James Brindley, which linked the duke of Bridgewater's coal mines in Worsley to the heart of Manchester.

Further south, where water power was less prolific, he would have spied the black smoke and belching steam from the hundreds of earth-wrenching Newcomen steam engines that were by now familiar sights the length and breadth of the country, pumping the mines free of water and driving the wheels of industry. Passing through the "black country" and the potteries north of Birmingham, Mason came across the excavations for the Trent & Mersey Canal, linking Josiah Wedgwood's pottery works at Etruria to the west coast, and passing through a tortured land of toxic waste dumps and new, mean housing that was springing up in narrow streets "thick with smoke even at midday."

Tired and weary, he finally arrived at his new house in Bisley, where Mary and the boys were waiting for him. By the time he had his quill sharpened and was writing to the astronomer royal with the results of his tiresome travels, it was October. He was able to assure Maskelyne, with some satisfaction, that he had found in the remote Scottish Highlands the perfect spot for testing the theory for the attraction of mountains—"a remarkable hill."

# 13

# IMPORTANT OBSERVATIONS

THE ROYAL SOCIETY'S SPECIAL COMMITTEE ON ATTRACTION, charged with managing the great project, included in its ranks Nevil Maskelyne, Daines Barrington, the Reverend Samuel Horsley, and of course Benjamin Franklin. So important was the experiment that later the committee was reinforced with the chemist Sir William Watson (1744–ca. 1825), Matthew Raper (1742–1826), and the botanist and future president of the society, Joseph Banks (1744–1820). Curiously, the scientist who should have been included was Henry Cavendish but for some reason (maybe because of his acute shyness) he was not.

On 27 January 1774, the council met in its chambers at Crane Court* and the business of the evening commenced. The fellows were pleased to note that the Committee on Attraction had received Mason's report of his journey through England and among the Highlands of "North Britain." Maskelyne read out the recommendations in full, concluding from that information that the Perthshire peak of "Sheehalian" offered a "remarkable hill" that seemed, to all intents and purposes, ideal for the great experiment. There was little doubt that the council would approve his recommendations.

On the astronomer royal's advice, the Committee on Attraction agreed that Mason was the "proper person" for the job and should conduct the observations. Certainly, Charles Mason and Nevil Maskelyne had already agreed on terms well in advance of the meeting (and probably before his exploration of Scotland), and the terms must have been acceptable to both parties. The next move, then, comes as something of a surprise. To conduct the experiment, the committee offered Mason a guinea per diem plus expenses.

Twenty-one shillings a day in 1774 was not an inconsiderable fee for, say, a half-decent lawyer, but for a scientific project such as the gravity experiment it was woefully inadequate. There were, for instances, the "inconveniences of the

---

* The Royal Society moved from Crane Court to nearby Somerset House in 1780.

climate" to consider and the fact that the "remarkable hill" lay at a remarkable distance from London, at the other end of the kingdom. Why the committee thought such a small fee was appropriate for such an important experiment is baffling. Of course, there were many other expenses to consider, such as local labor, the cost of modifying the zenith sector, logistics, and the purchase of a new theodolite, but there was more than sufficient cash available from what remained of the £4,000 granted by the king for the transit of Venus campaigns. There is also the possibility that Maskelyne hadn't actually expressed to Mason the figure he had in mind, rather just hinted at it; he was something of a tight-fisted, perhaps even miserly, man. And there is also the hint of suspicion that the committee thought perhaps that Mason, not being a university-educated man, was unworthy of a gentleman's fee.

Then there was also the question of who would assist Mason with the work. There would be a considerable amount of land surveying to be done, and on this point it is almost inconceivable that Mason would not have thought first of his old friend and very capable land surveyor, Jeremiah Dixon. Nothing remains in the archives to suggest that Dixon was consulted, but there again, the archives contain very little information; logic alone suggests his involvement. Whatever the truth of the matter, Dixon did not take part because he was busy in the family coal business and surveying the lands of the local gentry, he thought the whole scheme impossible, or like Mason, he was not in good health. At some point, probably shortly before or after the society's meeting of 27 January, it was announced that the assistant for the work was to be the mathematician Reuben Burrow (1747–1792).

When Charles Mason learned of the society's offer of a guinea a day, he turned it down flat. This came as a surprise to the council and posed something of a dilemma for Maskelyne. Certainly, the fee rightfully due Mason for conducting the experiment should have fairly reflected the great value that the Royal Society placed on the outcome. Even had Mason accepted the guinea, he would probably not have enjoyed the glory of proprietorship over the experiment; it was after all, a Nevil Maskelyne production. Poor Mason had been set up, and it is a reflection of the man's character that he refused to be cajoled into being Maskelyne's puppet. Perhaps it might be more appropriate to say Maskelyne's "fall guy" because there is just the hint that the astronomer royal harbored some small misgivings that the experiment might not actually succeed.

From this time onward we can sense Mason's increasing disenchantment with his lot. He was an extremely capable scientist and had done far more to deserve the recognition of his peers than many of the fellows of the Royal Society. He lived on an income that came principally from "piecework" for the Board of

Longitude, supervising the almanac computers and transcribing and much improving the lunar tables. He was probably augmenting this meager income by providing private tuition to the sons of the more wealthy Gloucestershire yeoman farmers. What is quite tragic, as will be seen later, is that he also missed an opportunity that could have radically changed his life. He had loved America and, ever since returning to Britain, had felt the allure of the West. At this critical time he must have thought longingly of what could await him back in Philadelphia, but when the latest news from America arrived, instead of following his heart, he prevaricated.

In the previous November, three tea ships had arrived in Boston harbor. Mass meetings of the populace had convened and there were calls for action. The angry townsfolk decided to order one of the ships, the *Dartmouth,* back to England without paying any of the import dues. Governor Hutchinson of Massachusetts reacted to the threat by ordering the authorities to impound the ship until the tea tax was paid. A few weeks later, a rally of Bostonians met to hear Samuel Adams report that the governor was standing by his word. That night, 16 December 1773, some of the Boston militants, masquerading as Indians, crept aboard the ship and raided the cargo. In total disregard for the correct way of making tea, the "Indians" stirred the contents of some 342 tea chests into the chilly waters of Boston Harbor. Parliament's response to the Boston Tea Party was to pass the first of the Coercive Acts (called the Intolerable Acts by Americans), shortly followed by the Boston Port Bill. The latter effectively shut the port to all commercial shipping until the taxes were paid and the East India Company reimbursed for the loss of its cargo.

This bad news apart, Mary Mason was pregnant again. Charles was obviously concerned for her health, given the tragic loss of his first wife. He need not have worried; on 11 April, he welcomed Robert, his fourth son, into the world. His wife survived the ordeal, as she would four more times, as well as outlive the astronomer. Shortly after the birth, Mason learned that Reuben Burrow, the Royal Observatory's "labourer," who had resigned some 5 months before, had accepted the job of performing the Schiehallion experiment. Burrow, a young mathematician of 27, was described as "an able mathematician but a most vulgar and scurrilous dog";[1] few of the scientific establishment liked him, but few could deny his competence. Even as Mason learned of Burrow's appointment, the man was on board ship with a mass of scientific paraphernalia, tents, and stores, heading north toward the River Tay and the passage up to Perth.

Like James Cook, Reuben Burrow (1747–1792) was a Yorkshireman, born in Hoberley, near Shadwell, the son of a tenant farmer. The young Reuben had shown remarkable abilities with numbers, and a local schoolmaster, Mr.

Crooks, taught him mathematics. About the age of 18, Burrow walked all the way from Leeds to London to seek his fortune by way of a clerk's position with a London merchant. Within a short time he had established his own mathematical school, in the naval town of Portsmouth, teaching mathematics to the midshipmen preparing for their navigation examinations.

As his mathematical skills became known, so Burrow came to the notice of Nevil Maskelyne. In 1770, the astronomer royal offered him the position of "labourer" at the Greenwich Observatory. A difficult, argumentative, and rather course man, his relationships with the genteel and refined Maskelyne were inauspicious and his employment was short-lived. He resigned in 1772 to marry Anne Purvis, a poulterer's daughter from Leadenhall Street, and opened a mathematical school in Greenwich.

The experiment that Maskelyne proposed to conduct on Schiehallion demanded the greatest precision that could be achieved at the time, and the astronomer royal had arranged for the finest equipment available to be at his disposal. Reuben Burrow's baggage on board ship included the Royal Society's 10-foot (304-cm) Sisson zenith sector, complete with its plummet modification; a 12-inch (30-cm) radius brass quadrant made by John Bird for the 1769 transit of Venus (the same one used by Dixon at the North Cape); and the much-traveled Shelton regulator used by Maskelyne on Saint Helena and elsewhere and used by Mason and Dixon in America.

There was also a transit instrument, strictly called a "transit and equal altitude instrument," and a brand new 9-inch (22.5-cm) diameter theodolite made by the celebrated Jesse Ramsden. Two De Luc barometers constructed by Edward Nairne for finding the heights of the mountain were included, together with a 66-foot (20.1-m) Gunter's chain for measuring distances. The society's 5-foot brass calibration standard, engraved by Bird for the Mason-Dixon expedition in America, and a linen tape measure completed the inventory.[2]

The latitudes determined by the zenith sector were "apparent astronomical latitudes," that is, the latitude measured by the stars alone. Then, as today, the only *observable* latitude was astronomical latitude, the appellation "apparent" being applied to anything that could be observed; for example, "apparent solar time," was the time shown by a common sundial before it was adjusted for a host of special astronomical corrections. The important thing to note about apparent astronomical latitude is that it was, and is, totally dependent on the direction of gravity.

To find the true, unaffected difference in latitude between his two observatory sites on the mountain, Maskelyne proposed using land-surveying methods. This true ground distance would then be converted to a difference in latitude by using the reverse of the process that had revealed the shape of the earth

The John Shelton Astronomical regulator. © Museums of Scotland, Edinburgh.

in the first place. The true difference in latitude would then be compared with the apparent difference in latitude as measured from the stars by the sector. If Schiehallion exerted an attraction on the plumb line, this would be revealed as a small difference between the two measurements.

In his earlier paper to the Royal Society, Maskelyne had set out the three objectives necessary to meet the requirements of the experiment:

1. To find by celestial observations the apparent difference of latitude between the two stations chosen on the north and south sides of the hill.

2.  To find the distance between the parallels of latitude.
3.  To determine the figure and dimensions of the hill.

The zenith sector included of a 10-foot-long telescope suspended near the object lens from a pivot set at the top of a stout wooden frame. The plumb line, or plummet, which provided the vertical reference, was suspended from the top pivot point and bisected the reading scale. By the time Sisson's zenith sector was on its way to Scotland, it was already some 14 years old. Apart from having the new plummet arrangement, the instrument had been refurbished at John Bird's workshops in Court Gardens, off the Strand. A new scale arc (called in those days an "arch") was fitted, divided up by "a new and arbitrary division adapted to the method of continual bisection"[3] so that it extended 8.5 degrees on either side of the vertical, a total arc of 17 degrees.

The art of dividing circles still relied upon the craftsman's eye and skill. As a consequence, whole units were often impossible to divide; thus one revolution of the instrument's slow-motion micrometer screw, for making the fine readings, produced the cumbersome value of 41.56″ rather than a whole minute (60″). Nevertheless, the instrument was remarkably accurate and allowed the observer to make readings of the scale to fractions of a second of arc.

To operate the zenith sector, the observer had to either lie on the ground to peer through the eyepiece of the instrument or recline on a specially made couch. Maskelyne did not record his preferred method, although he did have a special couch installed to observe through the fixed zenith sector in the Royal Observatory. As Maskelyne by this date was already in his early 40s and getting stout, we can grant the great man some comfort and allow him a couch.

The latitude was determined by measuring the "zenith distances" of those stars that passed by overhead and within 8° of the zenith (the point in the night sky directly overhead). The direction of the zenith was coincident with the direction of the plumb line and hence the direction of gravity. As a star came into view, it was tracked by using the tangent slow-motion screw, and as soon as it crossed the "wires" in the sector's eyepiece the telescope was clamped and the angle read.

The positions of the stars in the night sky are given in a system synonymous to latitude and longitude. The "latitude" is called declination and is measured in degrees north or south of the celestial equator; "longitude" is called right ascension and is measured in hours from a special astronomical origin called the First Point of Aries. The etymologies of these terms are astrological and are explained in the appendix.

Depending on whether the star was north or south of the zenith, the subtended angle was either added or subtracted from the star's known declination to give the colatitude, that is, 90° minus the latitude. Each observation had then to be adjusted for various astronomical motions to derive the final latitude of the observer's position. To keep accurate time, the Royal Society had loaned its much traveled Shelton regulator.

The next twist in the strange business came on 5 May, when the Committee on Attraction requested the Royal Society's council to intervene on its behalf to persuade Nevil Maskelyne to observe the experiment personally. Was it that they had no faith in Burrow as a competent astronomer? The man had something of a reputation for being offensive, but was it sufficient cause for the committee's decision? Was it that Maskelyne had always doubted Burrow as a competent astronomer and wanted someone else in his stead?

It would seem that this was quite likely the case, but at the time of the strange decision it is difficult to accept that Maskelyne wanted to do the job himself. He would have learned from Mason's report that Schiehallion was not the place for a middle-aged, learned, and reverend gentleman to disport himself, and certainly not a place for one as stout as "our astronomer royal," as he referred to himself. It must be concluded that he wanted Mason back in the game but that the old adventurer had once again declined. If the work was to be done that year, and there was little choice, the reverend would just have to go and do it himself; and he didn't like the idea one little bit.

The astronomer royal held his appointment from the king, so before he could go gallivanting around the Scottish Highlands, he had to get his sovereign's permission. He delivered his letter to His Majesty personally; presenting a letter to the king was, after all, something a little more serious than delivering a few complaining lines to Mr. Mapson in Tetbury. The king graciously allowed the astronomer royal to take as long as he wanted to "complete my very important observations."[4]

About 15 June, Nevil Maskelyne prepared to leave London for the long trip north. He would be traveling in company with his "esteemed friend" General William Roy, who had been involved with the project for some years and was probably instrumental in designing many of the surveying methods. The two companions arrived at James Lind's house in Edinburgh around 23 June. Here Roy set up one of his De Luc barometers, just as Mason had the previous year, as a "base" for his barometric height measurements of the hills.

They spent a few days with the Linds, recovering from the arduous journey and catching up with the latest scientific gossip. On Wednesday, 29 June, the two distinguished gentlemen set off with their Scottish guides on the short journey along the foggy banks of the River Forth toward Castle Menzies, near Aberfeldy. There they stayed the night as guests of Lord Menzies, occupying the same rooms that Bonnie Prince Charlie had 30 years before when fleeing the duke of Cumberland. Times and allegiances had changed forever; the days of the Jacobites were well and truly behind the ambitious lairds of the new Scotland.

The next morning, bidding their aristocratic host farewell, the two men set out bright and early on the last stage of their journey over General Wade's military road toward Rannoch. That evening they met with Reuben Burrow and William Menzies, a local land surveyor "who had been recommended by some of the principal gentlemen of the country, as a person proper for the work."[5]

The next morning was 1 July, and the four men and their ghillies set off on horseback to inspect Mason's "remarkable hill." Burrow explained as they went along how he had already chosen the observatory sites and "prepared every thing ready for the beginning of Observation."[6] Their first objective that morning was certainly to climb the steep southern slope up to the site of the south observatory. Here they found the site Mason had spotted, the "parallelepiped tent, 15½ foot square and 17 feet high," already erected and with the zenith sector installed.

Shelton's clock and the transit instrument were set up in a square tent a few yards west of the main observatory, and Bird's 12-inch brass quadrant stood on a barrel within a "circular wall [of stone] 5 feet in diameter"[7] and fitted with a rotating conical roof. Also standing alongside the tents on the bleak and windy hillside was a wood and stone hut, or "bothie," for the comfort of the astronomer royal during his sojourn on the rocky mountain. The sites of the two observatories can still be made out on the flanks of the mountain—vague patches of ground cleared of stones, each about 15 yards long and 5 yards broad, containing the faint ruins of low granite walls half buried in the tough grass and wild heather.

No records remain of where the scientists lodged; Maskelyne had his hut but it is almost inconceivable that he made it his permanent abode on the wild and stormy mountain. Then there were the two surveyors, Burrow and Menzies, plus at least four locally employed laborers to assist them. The celebrated fiddler Donnaeha Ruadh, Duncan "Redhair" Robertson of Rannoch, was engaged as cook and to help with all the entertaining that would go on, as well as to cater for the astronomers' creature comforts. There were riding horses for the scientists and packhorses for the equipment and stores and, possibly, a small wagon, all of which would have needed the attention of several men. In all probability, the experiment's team numbered a dozen or more men at any one time, all scurrying around the Scottish peak.

Remains of the observatory on the north side of Schiehallion, 1962. © M. J. Smyth, Institute for Astronomy, University of Edinburgh.

Maskelyne's "bothie" must have been a substantial building, capable of housing a considerable company, but it is highly improbable that Maskelyne would have entertained the thought of having common laboring men sharing his abode. Perhaps he would have tolerated his surveyors bedding down in his hut but none other; so it can only be assumed that the men, when they stayed on the mountain, must have made themselves comfortable in one of the tents. One possible candidate for a casual but more comfortable lodging and temporary headquarters is the mid-eighteenth-century mansion of Crossmount, home of the laird David Stewart, which would have afforded good cheer and a relatively easy ride to the hill. It is shown prominently on the map that accompanied the experiment's report.

In planning the experiment, Maskelyne was fortunate enough to be guided by two of the most experienced scientific field men of the age: William Roy and Charles Mason. The latter, who knew what would be involved in working on the mountain, provided advice on running a large operation, conducting the linear measurements and performing astronomy from a tent. Roy's military surveying experience would have been invaluable in arranging the survey team logistics and their working procedures; in his mapping of the Highlands, Roy furnished each of his surveyors with a noncommissioned officer and six private soldiers.

Schiehallion, the guidebooks tell us, means the "Fairy Hill of the Caledonians" and was locally also known from its rather obvious shape as Maiden's Pap. Maskelyne, in his paper to the Royal Society, refers to the name's meaning as, "in the Erse language, Constant Storm." In this he is clearly incorrect or was misled. It is a little surprising that he was not made aware of its correct translation during his stay, but the description was more than apt and the mountain then, as now, was famous for its bad weather and for burying its lofty peak in clouds for days on end.

Saturday, 2 July, was spent inspecting the sites for the two baselines upon which the land survey would depend and which Burrow had already identified. One of these baselines, named Rannoch, lay in a flat, boggy piece of ground northwest of the mountain and close by Crossmount. General Roy immediately saw the difficulties this posed, as the "ground for the Base was so very unfavourable."[8] That night, as they sat in front of the fire in their lodgings, sipping claret, Roy advised Maskelyne not to rely on the baselines but, instead, to "level and measure quite across it [the Schiehallion ridge] between the two stations [south and north observatories]."[9]

However, Maskelyne still needed a land survey from which to determine the volume of the mountain and to calculate its density. Despite Roy's reservations, Maskelyne also believed that the survey could provide him with the all-important distance between the two observatories. The scheme proposed for the land survey was extremely ambitious for the day and would result in the most complex trigonometrical scheme ever observed. Unfortunately, as we shall see, he neither appreciated the complexities of the task nor knew how it would be computed. However, he was developing a solution to the latter difficulty and had already identified the mathematician who would solve the problem.

Reuben Burrow, in his letter to the Royal Society the following year, claimed he had designed the survey plan and advocated its method of execution, and he may well have been responsible; he was very talented. A study of the progress of the work reveals a method sound in principle but poor in planning and less than perfect in execution. It would seem, looking back over the centuries, that in all probability the proposed survey method was the result of many meetings and discussions that took place in London, where Roy's influence both as a famous surveyor and friend of the astronomer royal would have been considerable.

Whereas the national and scientific mapping in Britain had not progressed as its many proponents had hoped, the "art" of surveying the land had im-

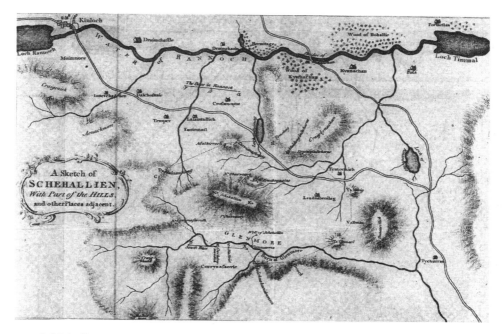

Schiehallion—map drawn to illustrate Hutton's paper in *Philosophical Transactions of the Royal Society* 68 (1778).

proved considerably, thanks in great measure to the promotion by the gentry of the English land divisions known as the "Enclosures" and to the "new engineering" demands of the Industrial Revolution.

Enclosing land was not something new or special to the eighteenth century and can be dated back to the Middle Ages. In fact, much of southeastern England had never known the expansive great fields that dominated the counties of the Midlands, East Anglia, and central southern England. The enclosure movement, breaking up fields of 600 or 700 acres into a dozen smaller plots, was fueled by many factors—the revolution in agrarian practices that demanded more economical land use, the price of cereal crops, and the social pressures of industrialization.

Before 1740, the fields of many villages were enclosed by consent of the manifold landowners, great and small, agreeing in a private treaty to combine their field strips and then divide the land into small, regular field plots. However, when a number of the smaller landowners contested the idea of enclosure in opposition to the richer and more powerful landowners, an Act of Parliament was needed.

After 1750, parliamentary enclosure became the norm, and passing over its more insidious social aspects, it had the advantage of imposing a rigorous procedure upon the enclosers. Every field enclosure required legal documentation and certification overseen by independent commissioners, who in turn appointed surveyors to map the land. The most prolific periods of parliamentary enclosure occurred during the years 1760 to 1780, when some 900 acts were passed, and between 1793 and 1815, during the Napoleonic wars, when high cereal prices, caused by a series of bad harvests, encouraged farmers to enclose land in order to produce more food and earn greater profits. Land surveyors were never short of work.

The enclosure movement changed the old open aspect of England into the neat and regular field systems of today. It also created a race of professional tenant farmers, replacing the ancient caste of smallholders, squeezed out by the high rents that landlords could charge their tenants. The social fallout from enclosures was mixed: the romantics (mostly town dwellers) looked back toward some idyllic pastoral age, whereas the more pragmatic country folk looked forward to better times. Poets like Oliver Goldsmith (1728–1774) produced sublime pieces such as "The Deserted Village" (1770), bewailing the loss of an ancient life as its rustic "swains" fled to the Americas:

> E'en now the devastation is begun,
> And half the business of destruction done;
> E'en now, methinks, as pond'ring here I stand,
> I see the rural virtues leave the land:
> Down where yon anchoring vessel spreads the sail,
> That idly waiting flaps with ev'ry gale,
> Downward they move, a melancholy band,
> Pass from the shore, and darken all the strand.

In 1783, Reuben Burrow's friend George Crabbe (1754–1832) cast a nostalgic backward glance to the days of his own youth in "The Village" (1783):

> Where are the swains, who, daily labour done,
> With rural games play'd down the setting sun;
> Who struck with matchless force the bounding ball,
> Or made the pond'rous quoit obliquely fall.

He needn't have worried: in rural Gloucestershire, the swains were still kicking footballs about the village street in the early years of the twentieth century.

The enclosures were just one aspect of the changes taking place in rural England as a result of the agrarian revolution. The shadows of the old fields can still be seen in the modern English landscape as curious regular humps and troughs, remnants of the ancient ridge and furrow of the field strips.

Land surveyors had been making large-scale maps of enclosure land for thousands of years. The earliest on record are those that reinstated the field systems of ancient Egypt following the annual flooding of the Nile valley. The sixteenth and seventeenth centuries saw an explosion in practicing land surveyors and great improvements in the "military art" of artillery, where the surveyor's skill and methods were critical to accurate gun laying to hit a distant (even invisible) target. These skills had improved further with the increasing demands of the enclosure commissioners for "before" and "after" enclosure maps.

A review of the number of surveyors practicing in England and Wales in the period well illustrates this increased demand. As the quantity of surveyors increased, so, too, did their skills and the quality of their equipment. In London, instrument makers like Nairne, Adams, Ramsden, and Troughton were producing high-class instruments. Yet every major city throughout the land supported provincial clockmakers and telescope makers who were turning out copies of the new theodolites and circumferentors. In Glasgow, James Watt, of steam engine fame, began his extraordinary career by producing fine surveying and other optical instruments and was himself a noted surveyor.

As the enclosure commissioners went about their business of closing up the land, so engineers began the process of opening it up. The eighteenth century is also known as the age of canals, and canal mania became the precursor of the ubiquitous mode of rapid mass transit of the next century. Canalizing Britain, as elsewhere in the developing world, was a key factor in fueling the Industrial Revolution, allowing raw material to travel cheaply to the factory and the finished goods to flow to the coastal ports for export. In 1757, the Sankey Brook Navigation, later called Saint Helen's Canal, was opened, followed in 1761 by Brindley's famous Bridgewater Canal, with its 60-foot-long aqueduct across the River Irwell, hailed as "this wonder of the age."

Between 1772 and 1777, some twelve new canal projects were underway. The new shape of industrial Britain was being sculpted out of its "green and pleasant land." By midcentury, the populations of Birmingham, Manchester, and Sheffield had swollen to 30,000 or more people, living under appalling conditions. "Black" Barnsley was producing iron and metal ingots that the Trent & Mersey Canal fed down the River Severn to the "noxious effluvia" of Bristol. All these great works needed maps for their designs and new and improved methods for setting out the lines and levels necessary for the canal water to flow.

Land surveyors such as John Henshall, Brindley's father-in-law; and his son Hugh; Samuel Simcock; and William Jessop began to develop the new methods and techniques. At first, their work was not of great accuracy, and even at the latter end of the century Nicholas Brown, who conducted the survey for the Huddersfield Canal, was still using compass and chain. Nevertheless, these surveyors and the new "civil engineers" were, between them, developing all the skills necessary for the mighty works of engineering that marked the climax of the last years of the century.

In the first years of the nineteenth century, the great engineer Thomas Telford, inspecting additions to the Huddersfield Canal, was able to report that although "the season was totally unfit for proving the general line of direction over the mountain and 'dialling' [observing angles underground] and levelling below, but each end appears very direct in itself" and "they have been tried and found accurate by several properly qualified persons [surveyors]." David Whitehead, the land surveyor who checked the earlier work found an error of just "two feet deeper than the sill at the first lock"[10] Thirty years before, such accuracy over such long distances would have been virtually impossible.

Even Jeremiah Dixon's land-surveying brother, George (1731–1785), "dug a stretch of canal on the fell in 1766" to drain the family's Cockfield coal mine. The Dixons were getting rich; by this date, the northeast was supplying 35 percent of the nation's demand for coal to fuel the growing numbers of foundries and smelters of the heavier industrial development.

# 14

## SO GREAT A NOISE

R𝐄𝐔𝐁𝐄𝐍 B𝐔𝐑𝐑𝐎𝐖 𝐀𝐍𝐃 W𝐈𝐋𝐋𝐈𝐀𝐌 M𝐄𝐍𝐙𝐈𝐄𝐒 and, of course, William Roy directly benefited from the explosion in knowledge and the availability of precision instruments that occurred in the latter half of the eighteenth century. The art of surveying was joining with positional astronomy in determining earth's vital statistics, and the science we now call geodesy was born. This, coupled with the new advanced mapmaking skills of the cartographer, enabled for the first time a realistic depiction of the earth. The challenge of adapting land survey techniques for civil engineering developed awareness in its practitioners for the unprecedented levels of accuracy necessary for the new and ambitious "mighty works of man."

Reuben Burrow had taken with him to Scotland a new, state-of-the-art theodolite built especially for the Royal Society by Jesse Ramsden. Curiously, William Menzies had brought with him the instrument's twin, loaned to Nevil Maskelyne at his request by James Stuart Mackenzie, the lord privy seal of Scotland. How this strange coincidence of identical theodolites came about when such instruments were rare and expensive is speculation. Possibly Mackenzie required a good instrument himself and was persuaded (by James Lind, maybe) to purchase a Ramsden instrument and make it available for Maskelyne on Schiehallion.[1]

In the event, having two identical instruments was to prove a godsend. Mounted on a rigid tripod, the theodolites appear to have been of the most precise sort, for their day. They had engraved brass circles 9 inches (22.8 cm) across, capable of providing angular readings by estimation to 1/4 minute of arc. They were fitted with telescopes instead of plain sights and no doubt were equipped with powerful magnifying glasses for reading the angles from the brass circle and a compass for finding bearings from magnetic north.

Burrow and Menzies had a month's start before Roy and Maskelyne turned up at Schiehallion. Work had already begun on surveying the mountain with

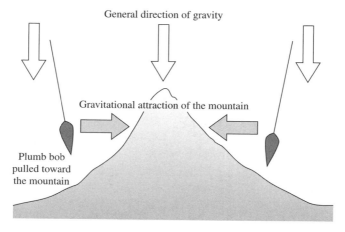

General direction of gravity

Gravitational attraction of the mountain

Plumb bob
pulled toward
the mountain

The Schiehallion gravity experiment.

the theodolites, and Burrow was optimistic that he could complete much of it; however, he had not reckoned with the terrible Scottish weather. Indeed the weather was atrocious: it rained hard and continuously for the next week, and little, if any, work was done at all. Maskelyne and Burrow entered the 17-foot-high "parallelepiped tent" (5.2 m) on 12 July 1774 to take the first stellar observations and begin the tedious business of determining the meridian (north-south line) to which they had to align the zenith sector preparatory to making the all-important latitude observations.

No sooner had they started work than the thick clouds and driving rain closed in again, ending any thought of astronomy. On Friday, 15, July, Roy completed his own contributions to the Schiehallion experiment. He had made many observations "both with his barometers & quadrant"[2] to determine the positions and heights of all the surrounding peaks. He then took his leave of the astronomer royal to make "some observations Geometrical as well as Barometrical . . . on the neighbouring Mountains"[3] of Strath Tay, as he explained in a letter to James Lind on 18 July. In the same letter, Roy noted that Nairne's barometers were performing well, although De Luc's method for ascertaining heights by barometer was "defective."

On Monday, 18 July, Maskelyne sat in his wood and stone house, imprisoned by thick clouds and driving rain. He had been on the mountain nearly 3 weeks, and it was time to inform his friend, James Lind, in Edinburgh, how they were progressing. Sharpening his quill, Maskelyne wrote, "From the Observatory, in the south side of Shihallien, July 18 1774": he complained that it had been a week since he first "saw a star in the Sector (so bad has the weather been)" but that he had been busy, with Burrow, getting the zenith sector aligned with the meridian. The astronomer royal was optimistic and, he noted cheerily, "that the clouds fly much higher than before," with the odd, rare patch of blue sky heralding a change for the better. Maskelyne estimated that he expected it would take 2 or 3 weeks before he could finish his observations on the south side of Schiehallion and another 2 weeks to transport all the equipment over the ridge to the north side.

Despite Roy's being "inclined to measure it by levelling directly across the hill," that is, measuring the distance between the two observatory sites with measuring "levels," Maskelyne informed Lind that he was determined to measure the distance by triangulation. He concluded his long letter with the news that the lord privy seal of Scotland was expected "every day at Castle Menzies," near Aberfeldy, and he finished with the express hope that it would not be long before Lind himself visited.

Meanwhile, Burrow and Menzies were busy with their land surveying. Their work was vital to the success of the experiment and had two objectives. First, the survey had to resolve the true ground distance between the two observatories, one on the south and one on the north side of the mountain. Second, the survey had to be sufficiently detailed to enable the hill's dimensions to be accurately determined, from which its volume could be calculated. Neither surveyor had ever attempted such a monumental task before; indeed, no one had ever tried to measure accurately the volume of a mountain, not even the illustrious William Roy. Although it can be argued that the survey "control" method was probably devised or recommended by Roy, the means employed to measure the mountain's three-dimensional shape was certainly the brainchild of Reuben Burrow.

Given the totally unique nature of their task and without any precedent to follow, it is not surprising that the two surveyors went about the business of surveying the mountain in a rather haphazard fashion. The arrival in our own time of satellite-aided land surveying has made the art and science of triangulation somewhat redundant. Yet for nearly 200 years, triangulation was the only means geodetic surveyors had for covering large tracts of ground. However, when Burrow and Menzies began their triangulation in the wet summer of 1774, the only scientific triangulation schemes ever attempted, with the exception of Cassini's French mapping scheme, were the long linear chains of triangles used to determine the distance of a meridian arc.

Burrow and Menzies decided to combine their triangulation work with observing profiles of the steep slopes of Schiehallion. Before beginning a survey, the first job was, and still is, to "recce" the ground. The surveyor would walk or

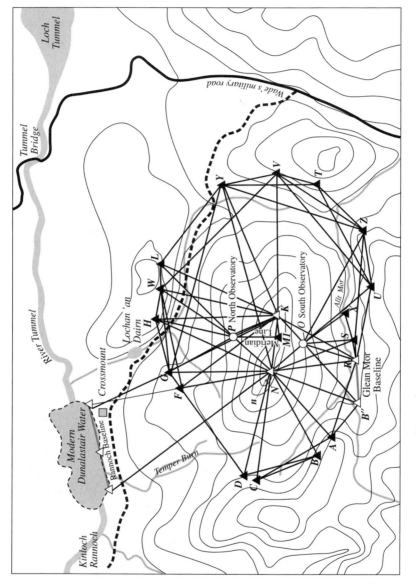

Burrow's trigonometrical survey scheme on Schiehallion.

ride over the land, noting the hills and valleys, the lines of sight, and the detail to be fixed, generally becoming acquainted with the nature of the ground. However, from examining the work of the Schiehallion team, it becomes apparent that only a rudimentary reconnaissance was made and was cursory in the extreme. Probably this had something to do with the terrible weather rather than inexperience and, perhaps, with the pressure exerted by the astronomer royal for the work to go ahead without delay.

Reuben Burrow was not an easy man with whom to get along. He made enemies of any who might disagree with him, referring to them as "scoundrels," "rogues," or "scallywags." He was bluff and outspoken and could lambaste a duke as quickly and as easily as he would a common laboring man. He was an insecure individual, heavily reliant on his few firm friends for comfort and reassurance; in short, he was mildly paranoid. In dealing with Maskelyne, Burrow found the astronomer royal's upper-middle-class background and privileged education an irritant, and he did not feel that Maskelyne deserved the position he held. Relationships between these two very different personalities became strained and would eventually break down completely as Burrow's persecution complex increased.

The two surveyors, as they circumnavigated the mountain's base, placed stout wooden poles on the rocky hillocks where they thought the survey stations should be. Little account was taken for the lines of sight or for which stations could be seen from what directions. This preliminary task done, the two men began to work from a convenient hillock on the southwest side of the mountain called Craignafarridh. This point they called $A$, and here they set up the theodolite and sent their laborers with arms full of wooden rods up the hill.

At a signal, the men stopped and placed the rods in the ground and the surveyors measured the horizontal and vertical angles to the rods. Then the surveyors advanced to the next point, $B$, and measured the horizontal and vertical angles to the same rod positions, using the first station as a "reference object," or RO. They soon realized that this was an inefficient way of progressing and decided to have the laborer hold the rod and to observe it simultaneously from two different stations with the two theodolites. This had the benefit of decreasing the amount of time the laborer had to spend planting and recovering the rods while increasing the rate of work. One surveyor kept the laborer on line and perpendicular to the contours of the slope, while the other measured the cutoff angle. In this way, profiles of the hill were observed, some 1,700 observations in all.

At each new station, horizontal and vertical angles for the triangulation scheme were measured to any station in sight, as well as to any additional stations that seemed appropriate or expeditious to include. This erratic means of progress is illustrated by the fact that, as they moved clockwise around the hill, more and more stations were added to the list in a chaotic, undisciplined way, with very little idea of the enormous problems that would arise when the triangulation scheme came to be computed. In addition, Maskelyne was making his own demands on the two men, and their patience, by adding his own surveying points to the list, which further tried Burrow's volatile temper.

The surveying work also had to fit in with the astronomer royal's demands for help with the nightly astronomical observations. At these times, Burrow would assist Maskelyne in the observatory while Menzies carried on with the survey as best he could. On Wednesday, 20 July, the two astronomers used their variation compass (a special compass designed to show true north ) to set the zenith sector with its "plane east"; that is, the alignment of the instrument was set north-south, with the scale facing to the east. They turned the instrument regularly to face either east or west to even out any errors in the alignment.

For nearly 3 weeks, Maskelyne had been trying to determine the direction of the meridian by using the quadrant and the Shelton regulator. This laborious task entailed observing the altitude of southern stars as they passed from east to west across the meridian of the observatory while simultaneously noting the time on the regulator clock as it clanked away the seconds. The time of transit, when a star crossed the meridian, was calculated by taking the mean of the time that the star was, say, at 40 degrees above the horizon to the east of south and then again when it was 40 degrees in altitude to the west side of south (a process called equal altitudes).

Astronomers always used local sidereal time, or star time, when observing because, unlike mean time, it could be calculated from the useful fact that the right ascension of a star and local sidereal time are the same when the star is on the meridian. By repeatedly timing this event over a period and using a combination of different stars, Maskelyne was able to determine the rate by which his pendulum clock was gaining or losing. Knowing this clock rate, he was able to calculate exactly when a star would be due south the following evening, and by observing the star at the precise moment, he could set out the line of the meridian and mark it with a pole in the ground.

It wasn't until 26 July 1774 that Maskelyne "drew a meridian" on the garden mats that covered the floor of the observatory and began the business of measuring the latitude. Unlike work in the relative comfort of the Royal Observatory, the observing process that Maskelyne adopted in his drafty, damp tent had to account for the lack of facilities. Burrow recorded the numbers and time while Maskelyne lay on his couch, peering through the telescope and making the fine adjustments on the tangent screw. Writing a year later, Burrow recorded that it was he who had "put up all the Instruments and Drawn the meridian line and put the instrument in order when he [Maskelyne] had put it out of order & did not know how to put it right again &c."[4] In this he was almost certainly recording the facts of the matter, and Maskelyne's failure to mention Burrow's part in the work severely irked the Yorkshireman.

The day they drew the meridian, William Roy, residing at home in Lanark, received a letter from Maskelyne with news of progress; he dutifully passed the information on to James Lind in Edinburgh. The weather had been poor, and the following week was no better. Cloud, heavy rain, and the occasional clear patch permitted just twelve star observations. The night of 1 August, however, was stupendous, and the great man must have been ecstatic, achieving, as he did, some 43 faultless observations for latitude with the zenith sector. Three days later, he managed another twenty-seven star sights before the rain and flying clouds returned to the mountain. Over the next 5 nights, he was able to collect just eighteen observations. Nevertheless, by 15 August, Maskelyne was satisfied with the number of observations he had collected and declared that work at the southern observatory was finished.

Before they could pack the equipment and move it to the north side of the mountain, Maskelyne required the position of the southern observatory to be connected into Burrow's burgeoning survey scheme. To do this he had the laborers build two large stone cairns on the top of the mountain, *N* at the western end of the ridge and *K* at the east. As the surveyors worked their way around the mountain, they took sightings on the cairns to incorporate them into the triangulation scheme. In this way, Maskelyne intended to calculate the distance between the cairns and use it as a "secondary" baseline to determine the distance (by more triangulation) from the southern observatory to the new site on the north side.

For some reason or other, it never occurred to anyone to actually measure the length of the secondary base, which, although arduous, would have been more than feasible and would have provided a more accurate result; perhaps Burrow did realize this but was being obstinate. As it was, the cairns were not visible from the observatory, and this was a severe difficulty. The problem was solved by another cunning stroke of lateral thinking by Reuben Burrow, which Maskelyne had to concede was very clever. Burrow knew his Euclidean geometry.

Crossing to the low hills on the south side of Gleann Mór, Burrow found a spot where the western cairn appeared in line with the observatory, and here he set a pole in the ground. He then repeated the same for the eastern cairn. Clambering back up the steep slopes to the observatory, he then measured the angle between the two poles with his theodolite. He then went up to the mountain-

Remains of the cairn named N. The white stone block is a modern triangulation pillar. © M. J. Smyth, Institute for Astronomy, University of Edinburgh.

top cairns and measured the angle from the opposite cairn to the two poles. Burrow now had a quadrilateral with two diagonals. Knowing that the opposite angles of two intersecting lines were the same, he was able to deduce his missing angle and complete a triangle that comprised the two cairns and the observatory. As a check, he added up the three angles and found they contained an error of just 2 minutes of arc. When the distance between the cairns was eventually worked out, Maskelyne could, by simple trigonometry, calculate the position of the observatory. The last thing Maskelyne demanded of his overworked surveyors, before they packed up, was to set up three poles to mark the line of the meridian that passed through the observatory and to include them in the exploding survey scheme.

On 17 August, the weary business of transporting the observatory instruments, tents, and stores to the far side of the mountain began. Already a gang of twelve laborers had been busy preparing the ground "with great labour" on the north side of the mountain, clearing a flat plateau among the rocks and the heather. This work occupied the next 10 days, during which time Burrow and Menzies continued their surveying of the hill.

On 26 August, the preparation for the observatory site on the north side of the mountain was complete. The weather, until then just poor, suddenly turned awful. On 1 September Maskelyne wrote a long letter to Roy at his Lanark home. "Nothing is yet fixed with respect to the distance of the two stations," the astronomer complained, referring to the mountaintop cairns. Maskelyne was beginning to doubt the wisdom of his idea of using triangulation and seriously reconsidered Roy's idea of a direct measurement between the two observatory sites. He requested that Roy should send him his "Telescopic Spirit Level," to which request Roy, writing to Lind, said he would comply and asked Lind to make the arrangements.[5]

In between the torrential downpours, rushing fog, and fierce winds, the laborers who reerected the observatory, built a new wood and stone bothie for Maskelyne's comfort. Burrow and Maskelyne set up the equipment, and finally, on the night of 4 September, the weather cleared sufficiently for work to begin. Maskelyne's first task was to collect star observations for setting the zenith sector into the meridian of the new observatory. Heavy rain again stopped play, and it wasn't until the wonderfully clear night of 7 September that they were able to prove the meridian alignment properly. That night they had the pleasure of observing forty stars for latitude. The next evening they only managed five and, tediously, nothing more until the middle of the month.

The morning of 16 September, Burrow and Menzies began the measurement of the baseline in the narrow Gleann Mór to the southwest of Schiehallion alongside Allt Mór's rushing waters, swollen by the streams that cascaded down the side of the mountain. Two measuring poles, each 20 feet long in the form of a tube, 3 inches square, and constructed of straight-grained seasoned pine, had been made by the local carpenter. These poles were carefully and repeatedly calibrated against Bird's brass standard until Burrow, who could be as fastidious as Maskelyne when he wanted to be, was satisfied with their accuracy.

The poles were supported on large wooden stands with triangular feet held firm in the soggy ground by iron pins. By means of slides on the stands and with the aid of a builder's spirit level, the poles were kept perfectly horizontal. Instead of trying to bring the "levels," as these measuring contraptions were known, together end to end, the men only brought them close enough to allow an engraved brass rectangular scale to measure the small gap. In America Mason and Dixon had, on Maskelyne's advice, brought the ends together either butt to butt or by means of a plumb line. It had been tedious work, and the method used by Burrow is certainly the result of Mason's experience and advice.

To keep the levels in a straight line, poles were set out along the line of the base, using the theodolite. After several days of work, Burrow was able to report to the astronomer that the baseline was 3,012 feet (918 m) from one end to the other. However, the men had also to report that the stands supporting the levels were shoddy and the ground very boggy. They were concerned that the accuracy of their measurement could not be assured. Consequently, new, improved stands were ordered.

The weather remained awful, and the survey was well behind schedule. Maskelyne's observations also suffered for the same reason, and it wasn't until 22 September that he was able to turn the plane of the sector to face west. Another line of meridian poles was set out across the hilltop on the 24th, just before the rain and fog returned. Nothing in the way of astronomy was done until a slight improvement in the weather on 2 October, when the astronomer managed eleven stars' observations. The next night was much better, and he had the pleasure of recording the passage across the meridian, near the zenith, of no fewer than fifty-eight stars.

On 8 October, Burrow and Menzies began measuring the Rannoch baseline on the north side of the mountain between the two points they called α and β This piece of ground is now submerged beneath the waves of Dunalastair Water; in 1774, it was a waterlogged marsh. Instead of using their wooden levels, the surveyors resorted to their trusty steel Gunter's chain,[6] which they dragged across the rough grass. The distance was measured at 4,184.73 feet (1,275.5 m) before correcting for temperature variations (later reduced to 4,182 feet by the mathematician Charles Hutton). A few days later, the distance was measured again, the levels being supported on new, improved stands. The wooden tubes gave a distance of 4,184.78 feet (corrected by Hutton to 4,183.56 ft). Finally, the entire baseline was measured to its terminal point, γ, and was found to be 5,897.12 feet (1,797.4 m). To be certain that all was done to the utmost perfection, Maskelyne personally supervised the entire miserable business. The weather throughout the measuring process had been terrible: driving rain and wind, exacerbated by an air temperature of 40°F (3.2°C).

On 19 October, Burrow was pleased, and relieved, to report to Maskelyne that the profile sections of the hill were complete. Five days later, when Maskelyne completed his own astronomical observations, the temperature at the bottom of the mountain had fallen to 38°F (3°C). With the onset of winter, the temperature continued to fall and the weather continued to deliver thick clouds and driving rain.

Throughout the "so great a noise" that the "uncommon experiment" had engendered among the great and the good of Scotland, Maskelyne was frequently visited by other scientists and the "neighbouring gentlemen." James Stewart (Stuart) Mackenzie, the lord privy seal, was one guest, as was his brother-in-law, Sir Robert Menzies, who assisted the experiment by arranging for the supply of laborers and materials and keeping Maskelyne well provisioned with "personal civilities."[7]

Alexander Wilson, professor of astronomy at Glasgow, and his university colleagues Thomas Reid and John Anderson were also visitors. Others included Lord Polwarth; Mr. Ramsey, the commissioner of customs at Edinburgh; the geologist John Playfair, who collected samples of Schiehallion's rock; Patrick Copland; and the Reverend Brice, who had ridden down Wade's military roads from the university at Aberdeen. On these occasions, the astronomer royal would entertain his guests in the bothie and show them around the observatory so they could inspect the cutting-edge technology firsthand. It is curious that James Lind was not mentioned as a visitor; perhaps, being a friend, his attendance, like Roy's, did not warrant comment.

With the coming of snow and hard frosts, it was time to abandon Schiehallion to its fairies and head back to London. The scientific documents and equipment were packed away and dispatched to Perth, ready for the coastal packet to London. The men were paid off, and Burrow made up the accounts for the Royal Society: the astronomer royal's expenses came to £212 0s 11d with a further £209 5s 4½d for the labor, materials, wagons, wood for the bothies, and Menzies's surveying fees.

However, the story of Schiehallion was not quite finished. On the eve of their departure from the mountain, Maskelyne threw a farewell party in his bothie. All the local potentates and gentlemen of quality were invited. A great quantity of drink and refreshments were imported from Edinburgh and London, and a keg of whisky was brought up to the observatory from nearby Rannoch. The laborers, who had through fair weather and foul, assisted in the great work, even though they probably had no idea what it was all about, were included on the guest list.

As the quality mingled with their inferiors and men of science discoursed with cowherds and shepherds, the food was consumed and the drink was drunk. The local musical ghillie, Red Duncan Robertson, whom Maskelyne had employed to while away the grim evenings with his talent on the fiddle, entertained the guests with lively Scottish tunes and sad Jacobite melodies. At some point late in the proceedings, as the guests laughed and the drink flowed, someone must have emptied the burning embers from his clay pipe or upset the stove for, suddenly, the bothie was in flames.

The guests, by now happy and carefree from their libations, tumbled out of the astronomer's hut to watch the flames engulf the little structure. Amid the dismay and inebriated laughter, Robertson stood despondent, his beloved fiddle burning within the conflagration that had been the astronomer's home. Seeing the man distraught, Maskelyne approached him to enquire what the matter was. Robertson, pointing at the flames, explained the tragic fate of his precious violin. Maskelyne comforted the distressed ghillie: "Never mind, Duncan, when I get back to London I will seek you out a fiddle and send it to you."[8] Some months later Robertson received a package from London. Carefully unwrapping the thick layers of brown paper, he discovered a black, leather-bound violin case. The beautiful instrument within was clearly foreign, made from a mellow yellow wood, beautifully crafted. Its tone was sublime. Beneath the frets was the maker's name: Antonio Stradivarius, 1729.

# 15

# THE ATTRACTION OF MOUNTAINS

AFTER NEVIL MASKELYNE LEFT THE MOUNTAIN at the end of October, Reuben Burrow stayed on, despite the fast-deteriorating weather, to make observations with the magnetic dipping needle and to complete all the sketches and perspective drawings of the mountain. Maskelyne, therefore, was the first to arrive in the capital and was soon giving out to his wide circle of friends and acquaintances accounts of his victorious gravity experiment. Rueben Burrow barely featured.

The secretary of the Royal Society, anxious to secure all the original observations as soon as possible, especially the land surveying so that mathematician Charles Hutton could start work on the computing, began making enquiries about their whereabouts. Christmas came and went and still there was no sign of the precious notes. In early 1775, the secretary was somehow or other given the false impression that Burrow was deliberately keeping back his fields notes. On 26 January, Burrow wrote a long and detailed letter from his home in Greenwich, replying to his Royal Society masters' demand for an explanation. He began his letter by apologizing for taking so long to answer, explaining that he had been ill. With the letter, Burrow enclosed the accounts for all the men's wages together with his own list of expenses as the experiment's paymaster.

In carefully measured tones, the passion barely constrained, he informed the society of the facts of the Schiehallion gospel according to Rueben Burrow. In his bright, clear handwriting, he explained to the secretary that not only had he kept copious notes during the experiment but he had also done much of the calculation. All this information he would have been pleased to include with his letter "had not Mr Maskelyne thought proper to take them into his together with a great number of Observations made by me on the barometers and thermometers, which I kept no Copy of."[1] For some reason, perhaps from spite or from idle indifference, the astronomer royal had in a convenient oversight forgotten to inform the secretary of the Royal Society that he had had the notes all the time.

Meanwhile, Nevil Maskelyne's pretentious first-person claims for all the credit, forbearing great hardship in the cause of science, and of his ultimate triumph reached Burrow's jealous ears. The society's letter was just the catalyst Burrow needed to put the record straight. He explained that, whatever any others might claim to the contrary, it was he who had chosen the observatory sites and who had erected the observatory tents on the mountain. He explained that it was he, Reuben Burrow, who had installed and regulated the astronomical clock, who had set up and adjusted the zenith sector, and who had "prepared every thing ready for the beginning of Observation."[2] All this, he wrote, he did long before Maskelyne arrived on the scene. After all, he had been laboring under the impression that it was he who was to conduct the experiment.

Burrow went on to explain that he had sorted out the survey plan and devised its method of execution and had located the sites for the two baselines. Had the weather not been so terrible, he wrote, much of the work would have been completed before Maskelyne arrived. In no uncertain terms, he informed the society that it was he, not Maskelyne, who had done most of the observations with the astronomical quadrant, as well as regulating the clock preparatory to establishing the meridian line.

Burrow was on a roll: angry with the society for robbing him of fame, he was emphatic that it was his idea to survey the hill in sections, on which point there is no doubt. He had even volunteered to stay behind in Scotland with Menzies to complete the survey with the two theodolites, but "Mr Maskelyne" had insisted that the instrument, borrowed from Stewart McKenzie, should be returned to its owner. Why Maskelyne should have insisted on this was a mystery, but, said Burrow, without both instruments he could not possibly have continued with the work.

That relations between Burrow and the astronomer royal were approaching an all-time low explains the acrimonious exchanges that were to follow. Perhaps Maskelyne was not so enamored of Burrow's work as Burrow had hoped or expected he should have been. The survey had been conducted in something of a haphazard manner, and as Hutton would later reveal, the observations were not without some errors and irregularities.

How much of the blame for this, if any, can be laid on Burrow's incompetence or general lack of experience is a moot point. Certainly he was, at that point in his career, something of a rough diamond and not averse to bluntly expressing his opinion of others, including the astronomer royal. But in his defense, Maskelyne's demands for additional survey stations and interference with the measuring of the baselines had tested Burrow's patience to its limits. In his own mind, Burrow believed he was due much of the accolade for the success of the project. And he was right to think so, but against him was Nevil Maskelyne, astronomer royal and egotistical leading light of the Royal Society.

The genesis for all this unpleasantness is revealed when, writing many years later, Maskelyne recalled,

> The Royal Society, thinking that the person then actually employed by them [Burrow] in the surveying part of the business had been my Assistant could not be depended upon to complete the work and publish the result, or at least, that the world would not be satisfied therewith on account of his inferiority of education and situation in life, made a point with me to go there to make the direction of the experiment, which I did, not without reluctance, not out of any wish to depart from my own observatory to live on a bare mountain.[3]

Reuben Burrow had turned out to be a very useful and competent assistant, maybe even a Charles Mason in the making. But Burrow was Maskelyne's junior by 15 years and was considerably lower down the social scale. Burrow was a forthright character, and from the tenor of his letter to the Royal Society, it was clear his contretemps with Maskelyne on the mountain had been serious. Later, Burrow reappears in our story, when the character and undeniable skills of the man are revealed by less opinionated persons who held his mathematical and observational capabilities in the highest esteem. However, at the time Burrow began his vitriolic attacks on his old master, attacks of a far more violent sort were about to break out between Britain and its American subjects.

In January 1775, Benjamin Franklin was still living with his adoptive family in Craven Street, London, where he assiduously pursued his scientific experiments and remained active in the Royal Society, where he retained his seat on the Committee on Attraction. In 1773, he had irritated the government by publishing the first of a series of satirical pamphlets entitled *Rules by which a Great Empire May be Reduced to a Small One,* hoping they would engender some common sense in British policy while pointing out what might happen if Britain continued to ignore the state of its American dependencies. In fact, all Franklin achieved was to increase the wrath of the government and raise suspicions back home in America, all of which led to accusations of conspiracy.

In January 1774, he had been summoned to appear before the Privy Council. The government's unequivocal position on the escalating revolt in America was that "advocacy of independence was sedition." The lord president of the council, Lord Gower, even recommended "reducing the Americans to submission." So well known was Franklin that among the 35 lawyers, peers, and ministers

Benjamin Franklin.
Courtesy Central Intelligence
Agency Archives.

present at the meeting that day were Prime Minister Lord North, the archbishop
of Canterbury, the lord chief justice, Lord Shelburne, Lord Germain, and Edmund
Burke. The attorney-general, Alexander Wedderburn, launched into Franklin
with an hour-long tirade. Wedderburn "poured forth such a torrent of virulent
abuse on Dr Franklin as never before took place within the compass of my
knowledge of judicial proceedings, his reproaches appearing to me incompatible
with the principles of law, truth, justice, propriety, and humanity," wrote one
observer. Franklin thought the experience was like "bull-baiting," but what
really caused the aging diplomat such pain was being slandered as "a man of
three letters; homo trium literarum—thief!"[4]

The privy councilors thought the performance highly entertaining, but
Lord North was dismayed: "When the Councillors applauded the speech and
then threw out the petition, they took a stand that for all their laughter was no
laughing matter." Throughout the humiliating charade, Ben Franklin had remained stoically reserved, "dressed in a full dress suit of spotted Manchester vel-
vet, and stood conspicuously erect . . . his countenance as immovable as if his
features had been made of wood."[5]

Franklin had entered the Privy Council chamber an untiring worker for unity and peace; he departed an implacable enemy of Britain. As Horace Walpole later wrote, "The calm philosopher, without reply, Withdrew and gave his country liberty." The next day, Franklin received a note advising him that he was no longer the deputy postmaster-general of the colonies.

In March 1774, the government had passed the first of the Coercive Acts in a direct response to the uprisings in Massachusetts; it demanded the tax due to the East India Company and restitution of losses caused by the Boston Tea Party. Two months later, the citizens of Boston retaliated with a boycott of all British goods until their port was reopened for trade. The government's reply was to place Massachusetts under martial law and to send General Thomas Gage with four regiments of British troops to firmly enforce its will.

When the legislatures of New York and Philadelphia called on all colonists to come together to resist the home government's action, Lord North's administration reacted swiftly. More repressive measures were introduced that virtually ended any semblance of self-rule then existing in British America. The Quebec Act pushed the ill-defined borders of Canada into territory that rightly belonged to Massachusetts, Connecticut, and Virginia. The new Quartering Act of June 1774 forced the colonists to welcome British redcoats into their homes. In response, delegates representing all the eastern colonies except Georgia held a long meeting in Philadelphia to discuss the situation.

Calling itself the First Continental Congress, the patriotic delegates included Patrick Henry, George Washington, Samuel Adams, and John Hancock. On 17 September, the Congress published its opposition to the home government's Coercive Acts, stating, treacherously, that they were "not to be obeyed." The Congress voted to form militia units and issued its "Declaration and Resolves against the Acts," which asserted the colonists' rights to "life, liberty and property." In the face of rebellion and trade sanctions, increasing colonial anger, and outbreaks of violence, General Gage decided to take action. In a show of imperial strength, he seized Massachusetts's arsenal in Charlestown.

The new year witnessed the arrival of the harbingers of war. On 9 February 1775, Massachusetts was declared to be "in a state of rebellion," and in March, Patrick Henry famously declared, "Give me liberty or give me death!" Franklin's quest had failed, and his cherished dream of "a mutually beneficial union of America and Britain"[6] disappeared in the smoke of cannon fire. When he learned of the events back home, he knew it was all over. On 25 March, he slipped quietly away from England's shores and back to America.

King George, ill advised and totally oblivious to his American subjects' resolve, signed the New England Restraining Act, which demanded that the colonies trade exclusively with Britain. General Gage received new orders to implement the Coercive Acts and meet the "open rebellion" with all necessary force. The resulting standoff, on Lexington Green, ended with the infamous "shot heard round the world,"[7] which marked the beginning of the American Revolutionary War.

The first months of the war saw the success of the American forces that were besieging Boston and captured Fort Ticonderoga on Lake Champlain. The Continental Congress met at Philadelphia and appointed George Washington commander-in-chief of the Continental army. On 17 June came the first major clash between the British and American troops, at the battle of Bunker Hill, on the north shore opposite Boston, where Israel Putnam famously ordered, "Don't one of you fire until you see the whites of their eyes."[8]

Such was the troubled news from the colonies when, on 6 July, Nevil Maskelyne stood to deliver the report on his Scottish experiment to his comrades of the Royal Society. Entitled "An Account of Observations made on the Mountain Schehallien for finding its Attraction," Maskelyne's paper was a long one. Its contents and the delivery left none of his audience in any doubt over whose idea it was in the first place, who was in charge, and who did the work.

Burrow's efforts on the survey were noted very briefly, although a good cough would have obliterated them. Roy's and Mason's contributions, which cannot be denied, were completely ignored; indeed, Maskelyne acknowledged only that Roy accompanied him as far as Edinburgh. There is no mention of Roy's physical assistance on the mountain or of the loan of his precious spirit level or of the logistical and scientific support given by James Lind.

In his address, Maskelyne was able to report a number of exciting preliminary conclusions. First, the true ground distance separating the observatories was calculated to be 4,364.44 feet (1,330.3 m). When converted to seconds of arc, using Bouguer's tables for the figure of the earth, this amount translated as a difference in true latitude of 42.94 arc seconds. Next, he revealed the actual observed astronomical results and showed that, by the stars alone, the difference in latitude between the two observatories was 54.6 arc seconds. By subtracting the former from the latter, Maskelyne proclaimed that "the sum of contrary attractions of the hill" was 11.66 arc seconds.[9] If one divided by two, to average the difference between the two observatories, Schiehallion was shown to deflect the plumb line by the tiny but measurable and real amount of 5.8 arc seconds, the equivalent of about 600 feet (183 m) over the ground.

From peak to base, Schiehallion stands some 3,000 feet and is, by any definition, a mountain. But it is not a very large mountain, nor is it particularly extensive. It is, as mountains go, beautiful but insignificant. Yet Schiehallion's mass was sufficient to cause a significant error in latitude, an error that was about twice what a well-calibrated chronometer would make in measuring longitude. The significance of the discovery for field astronomy and accurate map-making was stunning. Maskelyne cautioned future observers to "chuse those places . . . where the irregular attractions of the elevated parts may be small, or in some measure compensate one another." The alternative was, he mused, "to make allowance for the effect, which cannot but be a work of great difficulty,"[10] as indeed it was.

Today, surveyors have the benefit of a great store of knowledge about the shape of the earth and have maps and databases of the variable gravity anomalies, much of it derived from satellite observations. Yet still the work goes on, measuring and mapping the variations, so important is the effect on accurate geographical positioning.

Maskelyne was feeling pleased with himself as he summed up his other conclusions from the experiment. Sir Isaac Newton's "inverse square law for the gravitational attraction between bodies" was shown to be correct, and the earth's mean density was revealed to be at least twice that which it was at the surface of the planet, "contrary to the hypothesis of some naturalists, who suppose the earth to be only a great hollow shell of matter," a remark that, no doubt, drew a few chuckles from the astronomers and scowls from Joseph Banks and the "naturalists."[11]

Winding up his lecture, Maskelyne finished patriotically by acknowledging that the whole enterprise was achieved because of the "learned zeal of the Royal Society" and the "munificence of George the Third." The next day, the council of the Royal Society assembled, with one of their number missing. Nevil Maskelyne was requested to remain out of the council chamber; he would have been a remarkably stupid astronomer royal if he didn't realize what was going on. Sir John Pringle, the president of the Royal Society, proposed to the council that Maskelyne receive the society's highest award, the Copley Medal. The rules that stipulated that a serving member of the council was barred from the accolade were brushed aside: one of their own had triumphed. The fellows on the council concurred with the president, and the astronomer royal was up for another glittering occasion.

The award ceremony for the 1774 Copley Medal was held on 30 November 1775. In presenting the medal, Sir John Pringle noted that Maskelyne had endured "a residence of four months on the bleak mountain, and in a climate little favourable to celestial observations. To these inconveniences, however, he submitted with patience & complacency."[12] Perhaps Sir John should have consulted Reuben Burrow on this aspect of the great astronomer's temper. Winding up, the president was pleased to note that, thanks to Nevil Maskelyne, "the doctrine of universal gravitation so firmly established by this finishing step of analysis" could no longer be challenged. His congratulatory speech concluded, Sir John handed the precious prize to the portly astronomer.

Reuben Burrow did not share Pringle's or the Royal Society's vaulted opinions of the astronomer royal. In January 1776, he published a notice in the *St. James's Chronicle* announcing his intention of exposing "those who desire to have the Reputation and Appearance of being learned, without possessing any of the necessary Qualifications." He was very angry, concluding his advertisement, "At the same time will be published, Schehallien, a Poem, in the Stile of Ossian MacPhion; being a doleful Account of the perilous Perigrinations and dismal Disasters of the Sasnack Crean; intended as a proper supplement to the above."[13] James MacPherson published his contentious *The Poems of Ossian* in 1765, being, he claimed, translations from the Gaelic of the epic tales of the sons of Fingal (who supposedly lay slumbering beneath Schiehallion) "calculated to please persons of exquisite feelings of heart."

Burrow planned to use the "stile" of the poems to lampoon the "exquisite feelings" of Maskelyne and his friends. In the event, Burrow's promised prose did not appear in print. In all likelihood, the editor of the *Chronicle* thought it just too dangerous. In any case, Samuel Johnson maintained vehemently that MacPherson's translations were a fraud. When asked if he thought a contemporary man could have written the works, the wit replied, "Yes, Sir, many men, many women, and many children."

Whatever history may say about the character of Reuben Burrow, there can be no doubt that he did genuinely feel badly treated by the Royal Society and especially by Nevil Maskelyne, and with justification. But Reuben Burrow's days were by no means over, nor was the story of the Schiehallion gravity experiment at an end. Maskelyne is due the credit for devising and, with Burrow and Menzies, for undertaking the physical experiment and for his preliminary assessment of the results. But none of this could be described as new "science"; that work had yet to be done, and for doing it the society had chosen a self-educated mathematician from Newcastle by the name of Charles Hutton.

The soured relations between Burrow and Maskelyne were mirrored in the mounting gloom that pervaded London as more bad news arrived from America. It was becoming apparent to everyone that the popular sentiment at the start of the year for a quick war and happy resolution to Britain's differences with its truculent American subjects had been hopelessly optimistic. On 15 June 1775, William Roy wrote from his house in Argyll Street to James Lind in Edinburgh with the results of some atmospheric experiments he had conducted from the top of Saint Paul's cathedral, concluding, "American affairs look muddy; we may possibly have something else to do than observe with Barometers."[14]

Five months later, the general explained to Lind that he had "been for some time past extremely busy" on war matters. And again, on 13 January 1776, "You see then how matters stand; we are no doubt to have a Body of 17,000 Foreign Troops [German, or Hessian, mercenaries] in America next Campaign." On 22 February he wrote with some signs of hope: "There are authentick good news just arrived in 18 days from Boston. Montgomery had assaulted Quebeck in which he with about 90 of his men were killed. Arnold is wounded and taken prisoner with about 300 more. The remainder have retreated to Montreal. We have therefore reason to hope that Quebeck will be safe for the winter, and that now the Succours will arrive in sufficient time to secure it in the spring."[15]

The war had taken from William Roy what little hope remained for believing that his design for a national survey would receive a friendly hearing in government circles. All the mapping ideas were folded away as Britain prepared to dig in for a long struggle with its American colonies.

The day before Maskelyne gave his lecture to the Royal Society, the Continental Congress in America decided to try one last chance at reconciliation by dispatching Richard Penn, the erstwhile governor of Pennsylvania, to King George with an "Olive Branch Petition." They might have saved the boat fare: the arrogant monarch refused to either look at or discuss the petition, instead pronouncing the American people "to be in a state of open rebellion." Penn never returned to America. Marooned in London, he was not without a wide circle of friends, which included James Boswell and his famous anti-American mentor, Samuel Johnson. Boswell teased Johnson as frequently about his anti-colonial sentiments as Johnson taunted him about Scotland. On one occasion, wrote Boswell, Richard Penn observed Johnson prowling after the belle Elizabeth Hervey, following her up and down the drawing room. He overheard Lord Abingdon remark, "Your great friend is very fond of you; you can go no where without him." To which the lady responded, "Ay, he would follow me to any part of the world." The peer smiled wickedly, whispering, "Then ask him to go with you to America."[16]

The refusal of the king to have anything to do with the American petition and the lack of moral resolve within the government to think again about British policy led the rebellious American Congress to respond on its own—with the Navy Bill. In December 1775, the Continental Congress learned to its relief that France had given a favorable response to their pleas. Thomas Paine was moved to utter, with revolutionary zeal, "America shall make a stand, not for herself alone, but for the world." In March 1776, more bad news reached Britain; the Americans had evicted the British forces from Boston. Then, in May, Britain learned that France had delivered a huge arsenal of arms and munitions to the Continental army and that Spain had promised its support.

Amid all this war, the British Admiralty continued with its plans for exploration. On 25 June 1776, James Cook's two ships sailed from the Nore, bound for the Pacific and the West Coast of America. In a show of magnanimity, no doubt as a result of his happy days with the Royal Society, Benjamin Franklin secured a congressional guarantee that American forces would not attack Cook's little fleet. Sailing aboard the *Resolution* with Cook were seven American sailors and a 22-year-old sailing master by the name of William Bligh, a man with his own particular destiny.

As Cook headed for the American West Coast, General Howe was leading an army of 15,000 against Washington's forces on the East Coast. In August, Howe roundly defeated the America forces, precipitating a withdrawal to Harlem Heights. Another peace parley was held, but, again, the terms were impossible. After this last-ditch attempt at peace and face saving, Franklin and Silas Deane slipped away by sea to France to negotiate martial aid in their war with Britain.

British fortunes seemed to rally when news arrived in London that General John Burgoyne's army had landed in Canada and was on its way to join General Howe in a pincer movement around New York City. Plagued by poor communications and even poorer maps, Burgoyne finally arrived at the Hudson in early August, only to suffer a moderate defeat from the withering and sneaky fire of militiamen. Separated by the rough geography and unaided by any reliable maps, General Howe disembarked his troops on the Chesapeake and headed inland. In early September, he caught up with Washington at the battle of Brandywine Creek and forced the Americans back toward Philadelphia. The next month, October 1777, the two armies clashed at the battle of Saratoga, resulting in a major victory for the Americans. Burgoyne surrendered his entire army and was forced to take an oath "never to serve again in the war against America."[17]

News of the ignominious defeat was greeted with delight by Britain's enemies in Europe and with great despair by the Tory government of Lord North. Benjamin Franklin was honored by King Louis and informed that France recognized the true independence of America. Lord North offered his resignation to King George, only to have it turned down.

Many in Britain laid the fault for the war on the intransigence of the king rather than on North's administration. Echoes of Jacobitism were being heard in the drawing rooms of the well-to-do. Staunch Tory Samuel Johnson, in a fierce exchange with his Whig friend, John Taylor and in a reverse of earlier hopes, expressed the view of many in middle-class England on the unhappy state of affairs in the country: "If England were fairly polled, the present King would be sent away to-night, and his adherents hanged tomorrow."[18] In May 1778, General Howe was relieved by General Henry Clinton.

The Royal Society's Committee on Attraction had appointed the renowned mathematician Charles Hutton (1737–1823) to compute the results of Maskelyne's gravity experiment. From the volume of the hill and the deviations in the latitudes observed, Hutton was to derive the density of the earth and, hence, "weigh the world." Hutton was yet another example of the eighteenth-century's predilection for rescuing genius from obscurity.

Born the son of a Newcastle coal miner, Hutton began life alongside his father in Long Benton colliery. From an early age, he displayed great aptitude for learning and, through tenacity, became a schoolmaster in nearby Jesmond. In 1750, he started his own school for mathematics in Newcastle. His prowess with mathematics became well known through his book *Treatise on Mensuration* and in 1773 he become professor of mathematics at the prestigious Royal Military Academy at Woolwich.

On 21 May 1778, Charles Hutton presented his results for Maskelyne's gravity experiment to the fellows of the Royal Society.[19] Hutton's paper was exceedingly long and extremely detailed, reflecting the length of time it had taken him to solve the colossal problem. The paper was also modest; it was clear to all who knew him that the work was one of genius. "These calculations," wrote Hutton in something of an understatement, "were naturally and unavoidably long and tedious." His first task had been to abstract from Burrow and Maskelyne's notes all the angles they had observed. The original field books of the observations do not survive, but Hutton meticulously tabulated all the observations and included them in his paper.

His tables contain interesting comments on the quality of the observations, for example, when complaining about corresponding angles: "Such differences among corresponding angles I often meet with in the measures contained in the books of the survey, and it required much care to detect them, and trouble to reconcile them." And again, "There seems to be some general error in this section, as the depressions and altitudes are utterly incompatible with those of all the other neighbouring points in the plan." In abstracting the many triangles and checking their sums before calculation, he discovered a number of significant errors that caused him "much trouble to reconcile."

Hutton's next task was to adjust the lengths of the survey's two baselines, the one in Rannoch and the other in Gleann Mór, and correct them for the effects that temperature and humidity had on the measuring rods and chains.

Charles Hutton (1737–1823).

With a completed table of angles and the final distances for the two baselines, Hutton was ready to start computing "the great lines" that made up the many triangles that surrounded the hill. But, he grumbled "after great labour and pains," he had "frequently the mortification to find that the several values of the same line would differ so greatly one from another" that he could not rely on any of them. The problem for Hutton was that "the many small and unavoidable errors" that even the best theodolites of the day accumulated in the long chain of triangles, each dependent on its neighbor, grew to such an extent that it made the computers' task impossible.

After considering the problem, Hutton was forced to the same conclusion as Maskelyne, that the only solution was to compute the length of a "secondary base" between the two cairns at the east and west end of the Schiehallion ridge. From this secondary base, he proposed to calculate the lengths of all the other lines of the complicated scheme and combine them with the observations Maskelyne had made when determining his meridian directions. By doing this, Hutton was able to orientate the scheme to align with true north.

Starting with the southern baseline $B''$ to $R$ in Gleann Mór, the mathematician used the angles Burrow and Menzies had observed on the stormy mountain to solve each triangle, using tables of seven-figure logarithms until he had enough information to deduce the length of the secondary base $N$ to $K$. This, he calculated, was 4,052.2 feet (1,235 m). Repeating the tedious calculation from the northern baseline $\alpha$-$\beta$-$\gamma$ in the meadow of Rannoch, Hutton computed the same distance to be 6.7 feet (2.04 m) longer. Maskelyne, when he made his own calculations, had doubted the accuracy of the Gleann Mór baseline because of the swampy nature of the ground and unsteadiness of the equipment, and he had rejected it. However, Hutton was not persuaded to the same extent (probably after discussing the matter with Burrow) and decided to average the two lengths he had for the secondary baseline to give a value across the ridge of the mountain of 4,055.5 feet (1,236 m).

After computing the entire, massive triangulation scheme, Hutton then had to plot it all out. He constructed a map on stout paper, "four feet square," on which he plotted all the survey stations (there were about fifty), as well as the positions of the two observatories. This done, he next set about the incredibly tedious business of calculating "several hundreds of triangles conceived in a vertical position" to determine the heights of the thousand or more points the surveyors had observed during their profiling of the mountain. In all, seventy-two sections were observed from forty-seven stations, which made up combinations of several thousand individual triangles, each one solved by logarithms and a special slide rule that Hutton had devised. The height of each point was then meticulously plotted onto the map.

Hutton's next challenge was to calculate the volume of the mountain. His first inclination was to divide his map into some thousand vertical columns, "or pillars of matter into which the hill and the adjacent ground may be supposed to be divided by vertical planes, forming an imaginary group of vertical columns, something like a set of basaltine pillars, or like the cells in a piece of honey comb." The volume and attractive force of each column could then be computed separately and, added together, would provide a reasonably accurate result. Consulting his friend, the strange and reclusive mathematical genius Henry Cavendish (who had conceived the idea that oceans could have the opposite effects as mountains), Hutton worked out a better method for calculating the volume. Instead of dividing the hill into arbitrary cells, the solution they devised was to draw concentric circles around each observatory site and divide these circles by radial lines, like a sliced-up cake.

Hutton then realized that he had another difficulty to solve. Many of the "cells" of the cake did not fall where there was a surveyed height for a vertical dimension or else contained several different heights. After some cogitation, he thought of interpolating unit heights between the spot heights and "connecting together by a faint line all the points which were of the same relative altitude." In this one single clever piece of reasoning, Charles Hutton had invented contour lines, that most descriptive and distinctive feature of all topographical maps.

Now he could easily determine the heights from his contour lines of all the cells in the "cake." Because Hutton knew the dimension of each cell, it was simply a matter of hours of weary arithmetic to add them all together to come to a value for the volume of Schiehallion. From this, Hutton calculated that the gravitational attraction by volume of the hill compared to that of the whole earth was in the ratio of 87,522,720:8,811, or approximately 9,933:1, assuming that both Schiehallion and the earth were of equal density. This gave the theoretical attraction of the mountain as 20.8″ of arc. Now, Maskelyne had physically determined that the contrary attractions of the mountain amounted to 11.6 seconds of arc. The cotangent of this tiny angle is 17,781; therefore, reasoned Hutton, the mountain was denser than the earth as a whole in the ratio of 17,781:9,933, or approximately 9:5. In reaching this conclusion, Hutton could only assume that Schiehallion "consisted of an intire [sic] mass of solid rock," and he noted that the world would have to wait for a more thorough examination of the density of the mountain's rock.

"Common stone," said Hutton, was "2.5 times as dense as rain water," from which he concluded that the density of the earth was $9/5 \times 2.5 = 4.5$ times greater than that of water. His last line of reasoning was that, because the density of the whole earth was more than twice that of the mountain, "the interior parts" had to be considerably denser. With extraordinary insight, he predicted that the bulk of the earth comprised metals having densities ten times that of water.

Acclaim for his extraordinary labors was muted, eclipsing as they did the bright star of Maskelyne's achievements, and forces went to work to deny him recognition as the man who solved the Schiehallion enigma. Not until 43 years later, after Maskelyne was dead, would the truth emerge. Hutton justly deserved the Society's Copley Medal for his remarkable calculations and the discoveries that had emanated. He was fortunate in having already received it for his work on *The Force of Fired Gunpowder and the velocity of Cannon Balls* because he was very unlikely ever to be nominated again. The problem arose not with his mathematics but with the election of Joseph Banks to the presidency of the Royal Society. Banks detested Hutton and had accused him of neglecting his duties as the society's foreign secretary. The affair quickly degenerated into an acrimonious "Men of Science versus Macaronis" dispute.

# ~eeΟ 16 Ceee~

# THE BEST OF THE POSITION

ON 10 JULY 1778, FRANCE FORMALLY ENTERED the American conflict by declaring war against Britain. The whole dimension of the struggle for American independence changed: a family dispute was transmuted into another world war. The British generals fighting in America were disadvantaged by having to wage a major campaign far from home and without the prospect of rapid reinforcement. This desperate situation was made worse when France bolstered the unconventional backwoods fighting methods of the American militia with modern European-style warfare training for the Continental army.

Whereas Britain's generals in the New World were devoid of good or reliable maps of the vast territories of the American theatre of war, this was not a problem suffered by its fighting men on the other side of the world. The maps arriving in England of the immense land of Bengal were superb, thanks to the work of the first surveyor-general in India, Major James Rennell. Remembered as the father of modern geography, Rennell shares with Roy and Cassini the accolade of founder of modern map surveying. Regrettably, his star was to be eclipsed by other, more famous (though not necessarily greater) names.

James Rennell was not a well man when he arrived in England in February 1778. He was returning to his home country after serving some 10 years in Bengal in the service of the East India Company. His once strong constitution had been wrecked by years of working in the fever-ridden jungles and across the burning plains and from a series of terrible wounds that he had sustained. In Britain, Rennell's surveys in India were widely acclaimed and his homecoming was much welcomed, especially by William Roy and his old friend from India, Alexander Dalrymple.

The new president of the Royal Society, Sir Joseph Banks, was also pleased to have the geographer in the ranks. Rennell's fame had spread beyond the academic circles of the Royal Society and reached into the wider world even before he returned to Britain. In 1775, Richard Orme, Rennell's friend and an "eloquent historian of Hindostan," dined with James Boswell, where, referring to Rennell's achievements, he informed the young Scottish lawyer that "many parts of the East-Indies were better mapped than the Highlands of Scotland." Certainly, no one could doubt that. When Boswell related this story to Samuel Johnson, the wit replied in jest, "That a country may be mapped, it must be travelled over." Boswell, amused by the slight against his homeland, responded, "Nay, can't you say, it is not worth mapping?"[1]

In the realms of the mapmakers, the survey of India stands out as the cartographic jewel in the crown. It would come to symbolize the epitome of excellence in the formative years that lay ahead. Hundreds if not thousands of British and colonial surveyors would eventually come to trace their world-leading expertise to the methods developed by the acclaimed survey. However, the great subcontinent in 1778 was still many years away from being part of the "British Raj" of the Victorian and Edwardian era.

The British trading interest in India was made up essentially of a narrow tract of country along the eastern seaboard known as the Coromandel and the Circars, together with the vast province of Bengal, and was administered by the East India Company solely for the purpose of trade. The peace that followed the Treaty of Paris, ending the Seven Years' War in 1763, had led to a number of agreements between the belligerents. France had consented to restrict its interests in the Indian subcontinent to commerce and to recognize Britain's right to support the indigenous regional rulers from their headquarters at Madras and Calcutta. How this situation came about and how Britain founded a new empire to replace the one it was losing in America, as well as how all this had such a dramatic impact on the earth measurers, requires a little background information.

There is a saying, "Never read history backwards." This is sound advice and, in the case of India, mandatory if a fair and balanced picture of the country during the latter part of the eighteenth century is to be drawn. In the earliest days of Europe's interest in India, the northern half of the subcontinent was ruled almost entirely by the Islamic potentates of the Moghul Empire, whose cultural and ethnic roots lay to the northwest, across the mountains in Afghanistan. The three chief states of southern India were also controlled by Islamic lords; only in the extreme southeast were there independent Hindu principalities known as the Polygar.

Although the vast majority of India's population was Hindu, or at least not Islamic, the clash of religious culture was not a particularly difficult problem, and by and large Islamic rule of a Hindu people through the Hindu Brahmin caste of priests was relatively peaceful and relatively prosperous. This rule of

India by Afghani Moghul princes had been a near continuous feature of government ever since the first Moghul invasion during the eleventh century. That such a vast territory as India could be conquered and controlled by such a small occupation force seems at first incredible until one realizes that the cultures of India were perfectly designed for foreign subjugation. The rigid social structure of the Hindu caste system, exacerbated by the great variety of languages, effectively precluded the many castes, and even individual tribal communities, from joining in a concerted resistance against a determined invader.

The history of India, as it became apparent to Europeans, was in many respects similar to that of Scotland. The ruling classes were Mohammedans, similar in outlook to the Norman-French aristocracy of Britain. The percentage of Islamic settlers was very small, and the number of native peoples adopting Islam was tiny—with the exception of the provinces of Bengal and Punjab, where large numbers of low-caste Hindus had converted. In the tradition of Alexander the Great, the Moghul emperors divided their conquests into satrapies, each under a local lord called a *nawab* (deputy) assisted by a financial administrator, or *diwan,* whose job was to collect the taxes and tribute. When Portuguese, French, Dutch, and British traders first arrived in India, they obtained their trading licenses from the nawabs and their diwans.

In the latter half of the seventeenth century, the Moghul emperor, Aurangzib, conquered the southern half of India and created the region known as the Deccan. The consequence of his bold maneuvers was that he had spread his forces and his influence too thinly; his hold on the Deccan was shaky and his position in the traditional Moghul heartland of the north was weakened. Furthermore, and of more serious concern to the Moghul Empire, Aurangzib's conquests stirred up the Mahratta hornets nest. The Mahrattas were in many respects like the Scottish Highlanders, a tough, warlike people living by the warriors' code amid the hills and mountains of western India. Under their prince, Sivaji, the Mahrattas suddenly became a formidable center of power and began raiding deep into Moghul territory.

When Emperor Aurangzib died in 1707, he left behind something of a power vacuum. The nawabs of the Deccan, Bengal, and Oudh saw their opportunity for independence from Delhi and took it. Taking advantage of the weakening empire, the Mahrattas extended their raiding and conquests across India and exacted *chauth,* the counterpart of the Highlanders' "Black Meall," from local princes and the poor alike. Despite their strength in numbers, the Mahrattas lacked the administrative cohesiveness necessary to become an alternate empire. Instead, they satisfied themselves with plundering and with moderate land acquisitions and settlements in the Punjab.

In 1739, the ruler of Afghanistan and Persia, Nadir Shah, swept down through the Khyber Pass in a devastating raid on the Moghul capital of Delhi and virtually destroyed what was left of the Moghul Empire. Nadir Shah fought against the Mahrattas and wrenched the Punjab from them, establishing his own regime, before returning from whence he had come. In 1759, the Mahrattas retook the Punjab, which prompted Nadir Shah's successor, the Afghani Ahmid Shah, to retaliate with swift retribution. For 2 years the warriors of both camps raided and fought one another across the tortured land until, at the battle of Paniput, the Mahratta army was slaughtered. But, instead of consolidating his position, and installing satraps to rule in his stead, Ahmid Shah emulated his predecessor: he packed his trunks and went back to Kabul. However, the Mahrattas had been badly beaten, and their slow recovery never reached the status of their heyday.

While the native Asian potentates battled it out, the French and the British followed suit, but on a more modest scale. The trading establishments of both nations maintained small armies of well-trained native *sepoys* led by European officers. The French, as ever, seemed to be more astute with Indian politics than were the British and positively adept in getting on well with their Indian nawab hosts. In 1744, the War of the Austrian Succession (1740–1748) arrived in India. For 3 years, the two trading rivals maintained a sort of peaceful impartiality to the war, arguing that their purpose was trade and not national aggrandizement. However, François Dupleix, the governor of Pondicherry, and Labourdonnais, the governor of Mauritius, saw in the war an opportunity to rid India of their British rivals. The French prepared large land and naval forces, and the British government, suddenly perceiving the value of the East India Company, dispatched a naval squadron and a small army.

Anwar-ud-din, the nawab of the Carnatic, anxious not to have the Europeans embroiling his people in their squabbles, warned both trading houses not to wage war in his territory. In 1746, the French traders chose to ignore the ultimatum and attacked Madras, taking both town and Fort George. Anwar-ud-din, angered by the French duplicity, led a counterattack with an army of 10,000 men. Despite their vast superiority in numbers, the Indians were driven back. In 1748, it was the British fleet's turn to counterattack with a raid on the French base at Pondicherry, but it failed to make any impression on the stout defenses.

In the eyes of the local rulers, the French seemed invincible, but just when it looked as though the British were finished, news arrived of the Treaty of Aix-la-Chapelle, ending the war. Having lost, the terms imposed on France were to restore Madras to the British in exchange for the return of Louisburg, on the Saint Lawrence. However, the cunning Dupleix was not to be dissuaded from his goal

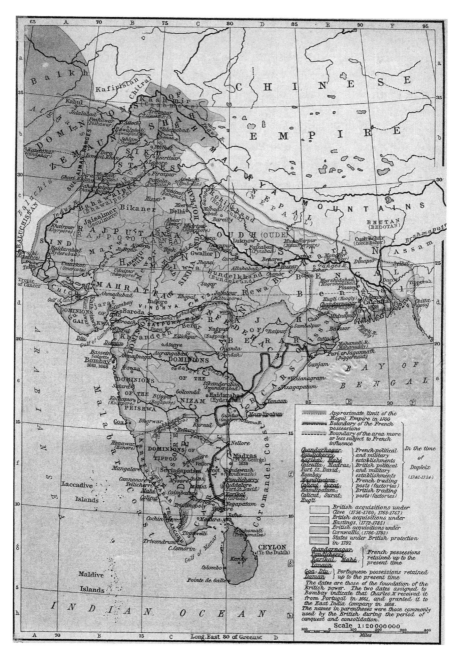

India at the time of Clive. Illustration from *The Historical Atlas* by
William R. Shepherd, 1923.

of dominating the coast. Instead of confronting the British head on, he decided to provide his considerable military and diplomatic support to one of the strongest nawabs of southern India, then at war with the local potentates.

The East India Company responded to Dupeix's strategy with one of their own and turned to the rulers of Bengal and the northeastern states. The French scheme worked well, and by 1750, Dupleix had installed puppet nawabs on the thrones of Hyderabad and the Carnatic. He had also built up a very considerable sepoy army, paid by the thankful nawabs and commanded by his own French officers. Anwar-ud-din, who had been so upset by France, bringing war to his land, was slain and his son, Mohammed Ali, was cornered and besieged in the southern city of Trichinopoly.

Mohammed Ali was the only obstacle standing in the way of total French control of Indo-European trade. All seemed hopeless to the beleaguered British traders of the northern Carnatic, but then a new governor arrived in Madras—Thomas Saunders. Saunders astutely weighed up the situation and sought the advice of Robert Clive, a young soldier who had distinguished himself during the fight at Cuddalore's Fort Saint David. Clive's plan was quite brilliant; instead of trying to take the superior French forces head on and loosing, he instead suggested attacking Arcot, the Carnatic's capital city.

Clive's assault on Arcot with just 500 men was a complete success and a famous victory. The French puppet nawab of the Carnatic, Chanda Sahib, reacted exactly as Clive anticipated and withdrew a large part of his forces, which were besieging Mohammed Ali at Trichinopoly, to march on Arcot, 200 miles to the north. Chanda's siege of the city's fortifications lasted 50 days before it was raised by Clive and his troops, sallying forth and defeating the nawab's 10,000-strong army. In Trichinopoly, Mohammed Ali, hearing of the victory at Arcot, drummed up support for his cause from other Indian rulers, including the fearsome Maratthas. When Clive and Lawrence came to his aid in June 1752, they joined their forces and overwhelmed the French. British hopes and aspirations seemed restored.

A local potentate executed Chanda Sahib, and Mohammed Ali was installed on the throne in his place. However, what seemed a promising situation for the East India Company was an illusion. The French were still numerically superior and had the support of the powerful nizam of Hyderabad, whom they protected from the feared Mahrattas. In return for this arrangement, the nizam ceded to the French the region known as the Northern Circars, between Madras and Bengal, an area nearly the size of England. This was the first Indian territorial possession of a European power. So it was that, in 1755, on the eve of the deciding conflict between France and Britain, "the French seemed to have all the best of the position, in India as in America."[2]

Robert Clive, during his brief return to England, urged the East India Company's directors to reinforce their tiny army to fend off the inevitable French attack. Clive's reputation and heroic status preceded him, and therefore it was impossible for either the company or the duke of Newcastle's administration to refuse his demands. A much larger and better equipped army was recruited and billeted in the company's great warehouses in London's Cutler Street, now refurbished as fashionable residences and boutiques. Apart from his new troops, when Clive returned to India in 1756 he had the aid of an entire squadron of Admiral Watson's warships. He also took with him his new wife, Margaret Maskelyne, none other than the sister of the astronomer royal.

News of the official start of hostilities of the Seven Years' War still hadn't reached Madras when news of a more terrible sort did. The nawab of Bengal, Siraj-ud-Dowlah, a bitterly twisted young man, had launched an attack on the company's factory and fort at Calcutta, allegedly because the company had refused to cease its fortification of the city in the face of a potential French assault. The small British garrison had been quickly overrun, but many of the leading residents and its president were able to escape. One of the members of the council, John Howell, together with a number of Europeans, was taken captive. According to the reports of the time, 146 men and women were locked in an airless cell for the night of 20 June 1756. Next morning, when the cell was opened, all but 23 had suffocated. More recent studies have suggested that the number of persons locked in the notorious Black Hole of Calcutta was 64 and that 21 survived. Whatever the truth of the matter, the incident drew the severest condemnations, and General Clive and Admiral Watson immediately determined to take their revenge and forcefully restore Calcutta to the company.

At the beginning of January 1757, the British forces were in position to retake the city. The nawab's troops marched out to meet them. The battle was a decisive victory for the well-trained and well-armed British forces, and the nawab was forced to sign a treaty. But Siraj-ud-Dowlah had no intentions of honoring its terms and almost immediately began negotiating with the French commander in nearby Chandernagore to launch another attack on the British.

News of the outbreak of formal hostilities between France and Britain, the Seven Years' War, arrived just in time to back up Siraj-ud-Dowlah's arguments for reprisals. When word arrived in the British camp that the French had dispatched a large army and naval task force under the command of the Irish exile Count Lally, Clive was spurred to action. He turned on the French fortress on the Houghli River while Watson bombarded the city of Chandernagore. The sudden and ferocious assault was followed by fierce fighting and the fall of the French strongholds. This bold maneuver had the desired result, temporarily relieving the pressure on Bengal.

Siraj-ud-Dowlah was a tyrant, and his oppressive rule had the effect of alienating many of his advisors and supporters. In desperation, they entered into a conspiracy to dethrone or kill him and install his father's old friend and *bakhshi,* Mir Jafar. The conspirators approached Clive for a deal, bringing with them a Hindu banker called Omichand as an intermediary. Suspecting a subterfuge, Clive drew up two agreements, one fake and the other genuine. The fake document mentioned Omichand's commission; the other did not. This slight for pecuniary gain would return to haunt Clive. Some modern historians regard this sneaky deal—to secure Mir Jafar's assurance that he would not support Siraj-ud-Dowlah—as Britain's way of securing India by bribery and corruption. Reading history backward is a fatal error; what Clive did, by the ethics of the day, was certainly questionable but did not win an empire. The secret treaty was signed in May; if Clive was successful, Mir Jafar would have the throne of Bengal and Clive and the honorable company would receive the equivalent of £3 million and the taxable rights to the country around Calcutta.

In June, Clive's 3,000 troops marched against Siraj-ud-Dowlah's capital of Murshidabad, 100 miles north of Calcutta. Opposing him were some 50,000 Indian fighters, including Mir Jafar's men. After some hesitation and, like Rob Roy at Sheriffmuir, uncertain of the capricious Mir Jafar's intentions, Clive decided he had no choice but to attack or lose the company's tenuous hold on Bengal. He ferried his men across the Houghli River and marched on the enemy's camp at Plassey. The first part of the battle was little more than posturing by both armies, each withdrawing to reassess their positions. Cannon fire was exchanged, and when one of the nawab's batteries withdrew from a forward position, a resourceful British officer and his gun crew quickly occupied its place, much to the irritation of Clive. But it proved an inspired move; the rapid fire of the forward British cannon crew removed any lingering doubts in Mir Jafar, and his men deserted. Siraj-ud-Dowlah's troops were so unnerved by the mass desertions and the precision of British artillery that they fled in panic. It has been estimated that Indian losses were fewer than 500 to just 19 British dead. Plassey was a famous victory but scarcely deserves to be called a battle, although its consequences were perhaps the most portentous that Britain had ever experienced. Siraj-ud-Dowlah was caught and executed by his own followers and Mir Jafar was installed as nawab, a puppet ruler completely dependent upon his British friends.

The French were by no means finished. The long-expected fleet carrying the French army reinforcements arrived at Pondicherry in June 1758; they still had a large army in Hyderabad. In December, the Hyderabad troops marched to join up with the forces from Pondicherry in an attack on the British factory and fort at Madras. For 2 long months, the French besieged the city before a British naval squadron finally came to its rescue.

The actions in India before and during the Seven Years' War brought to the East India Company vast tracts of country through treaty and agreements with the nawabs and by victory over the French. Unfortunately, the company's directors in London were reluctant to acknowledge their new protogovernmental responsibilities. Clive's reputation among the directors in London was not high: "You seem thoroughly possessed with military ideas as to forget your employers are merchants."[3] The problem was that, despite its enormous trading fortune, the East India Company was not financially strong. The rising costs of its private army and the huge administration, plus the level of taxes imposed on it by a hard-up British treasury, began to be too much for a private concern to bear. The East India Company found itself in possession of provinces as extensive as the combined area of France and England.

In 1759, Clive had written in secret to William Pitt that "so large a sovereignty may possibly be an object too extensive for a mercantile company; and it is to be found that they are not themselves able, without the nation's assistance, to maintain so wide a dominion."[4] concluding his letter by suggesting "the government's taking it in hand." As soon as Pitt became prime minister in 1766, he decided it was time to bring the company's vast territories within the sovereignty of the Crown. This first move toward ministerial control failed for many reasons; nevertheless Parliament was able to impose on the company certain obligations, including the importing of unwanted British woolen goods into India.

The company's "traders" in India were more learned in bookkeeping than administration, but even to them it was obvious that they were in possession of vast and uncharted lands. A lack of maps was becoming a serious problem. Even prosecuting the war had been hampered by the inadequacy or shear absence of maps. On more than one occasion, such as the march on Masulipatam, the deficiency of the mapping had nearly resulted in disaster. Despite the company's fiscal difficulties and other worries, it was at least able to begin redressing the problem of its lack of maps: its governor in Bengal had hired himself a surveyor.

# 17

## DISTINGUISHED MERIT

THE VICTORY AT PLASSEY IN 1757 had brought General Clive and his employers the vast territory of Bengal. The exact extent of the land area or the positions of its borders were somewhat nebulous. The few maps of the region that existed were sketchy in the extreme, made up of reports from company men who had penetrated the interior, a few astronomical observations, and the tales of Jesuit travelers. Clive was also conscious of the fact that during his military campaigns, his maps had been little better than hopeless. What he needed was someone to compile a gazetteer from the volumes of existing information about the country and to make new surveys.

The circumstances surrounding the appointment of James Rennell as the first surveyor-general in India were as unlikely as they were fortuitous. Between 1764 and 1777, Rennell directed an unparalleled feat of mapping that spanned the dominions of the East India Company as far north as the Himalayan Mountains. Considering the vastness of the areas covered, the difficulties he encountered, and the speed with which the work was accomplished, Rennell's mapping in India was a remarkable achievement that stood the test of time. Indeed, he is counted among the most able geographers of the day, and although ill health made it impossible for him to continue his practical survey work beyond 1777, he left his mark as geographer, antiquarian, and adviser to the Indian Survey Office for something like half a century.

James Rennell was born on 3 December 1742 in Chudleigh, Devon.[1] His father, an army officer, had been killed during the War of the Austrian Succession shortly before his son's birth. In 1756, at the age of 14, young Rennell joined the Royal Navy as a midshipman, serving under Captain Hyde-Parker of HMS *Brilliant,* and saw action in the Seven Years' War. He was diligent and a quick learner and mastered navigation at an early age. He first demonstrated his natural skill at mapping during the battle of Saint Cast Bay, where he made detailed plans and charts of the actions. When Hyde-Parker got command of the *Norfolk* (74 guns), bound for the East Indies, Rennell was invited to join the ship's company. Hurriedly returning from a spell of shore leave, Rennell's packet was wrecked in passage, and by the time he reached Portsmouth, the *Norfolk* had sailed.

Fortunately, he met Captain Haldane of the frigate HMS *America,* bound for India, and Haldane agreed to take Rennell on in hopes of catching up with the *Norfolk.* The *America* set sail on 6 March 1758 and reached Madras 6 months later, in September. On arriving in India, Rennell discovered that Hyde-Parker had had another change of command, this time to the *Grafton* (68 guns), a third-rate ship of the line. Rennell was soon back with his old shipmates and resumed his casual surveying projects.

In November 1761, he surveyed the coasts of Trincomali (Sri Lanka) and Rodriguez Island. In March 1762, his first 6 years of service with the Royal Navy was completed and he was entitled to apply for promotion to lieutenant, but as Hyde-Parker explained to him, the war was all but over and promotion unlikely. In April 1762, Hyde-Parker arranged for Rennell to be seconded to the East India Company to accompany the trader-cum-hydrographer Alexander Dalrymple on an expedition to the Philippines and to assist with drawing charts of the coasts. A friendship between the two men began almost immediately and was to last a lifetime. Rennell so enjoyed the work that, when he returned to Madras, he requested his discharge from the Royal Navy so that he could join the company's fleet.

In July 1763, he was given command of one of the company's 300-ton coastal traders and a salary of £300. Unfortunately, his first command did not last long: a hurricane in October wrecked his ship, along with most of the fleet, which was sheltering in Madras roads. Rennell had lost everything except his friends. One of these, Governor Robert Palk of Madras, found him the command of the *Neptune,* a merchantman running supplies to the troops that were besieging Madura. Taking advantage of a lull in the conflict, Rennell conducted some charting surveys around the southern tip of India, Cape Calimar, and among the many dangerous channels and reefs of the Indian Ocean. Returning to Fort Saint George, Madras, James Rennell suddenly found himself something of a hero, or at least a celebrity, and was rewarded with "a handsome present of money."[2]

From Madras, Rennell took the *Neptune* north to Bengal and up the channel to Calcutta. There he ran into an old navy friend, Captain Tinker, who introduced him to his friend Henry Vansittart, the governor of Bengal. Rennell's coastal surveying adventures were well known among the company's servants, and the governor was delighted to hear of his exploits firsthand. Vansittart was well aware of the problems the company suffered from having an inadequate and unreliable map library and was "anxious to inaugurate some system for . . .

correcting and revising the received geography of Bengal."[3] He was ever on the lookout for a likely man to take on the job of surveyor of Bengal, and Rennell seemed the perfect choice. Much to the young sailor's surprise and delight, Vansittart offered him the job on the spot. The official appointment as surveyor of Bengal came through in April 1764 (Rennell was appointed surveyor-general in 1767). A few days later, he received another pleasant surprise. Another old shipmate, Henry Topham, had secured him a commission as an ensign in the Bengal Engineers. James Rennell was just 21 years old.

His first season as surveyor of Bengal began at Fort William, Calcutta, where he organized his work plans and made a start on compiling the volumes of dusty documents that had been piling up over the years in the company store. Taking advantage of the coolest part of the season, he set out on a reconnaissance of the Ganges River, making astronomical observations with a quadrant as he went along. During his first season, Rennell lived aboard an old "budgerow," a primitive type of houseboat that was towed up the river as a sort of mobile headquarters.

Exactly how far Rennell ventured up the Ganges is a matter of debate, but he probably got no further than Patna. The Ganges, the great river of India, was known to rise from sources in the tall mountains to the north and flow across the plains of "Hindoostan," fed by many tributaries. The Ganges Plain is extremely flat and in those distant days was a patchwork of cultivation set between swathes of dense jungle and tall grasses. It was the home of elephants, leopards, and the ferocious Bengal tigers, as well as venomous snakes such as the fearsome 18-foot (5.5-m) hamadryad, which can raise itself to the height of a man and inject enough venom to kill an elephant.

Returning to Fort William in the fall, at the start of the rainy season, Rennell reviewed the information the governor had been collecting; it was an inauspicious pile. After examining the sketch maps and papers and perceiving their inadequacy, Rennell prepared a proposal to the council for correcting "the whole geography of Bengal." Governor Vansittart, about to depart for England, was delighted with the proposals and promised Rennell to "put your services in such a light to the Company that they may give you the encouragement that your diligence deserves."[4] For his "diligence," Rennell was rewarded with a salary of £1,000 a year.

On 14 January 1765, Rennell was promoted to lieutenant of engineers. That season, he began surveying a stretch of territory "2½ degrees of longitude and 3 of latitude," an area of some 38,000 square miles (100,000 km²). This work brought him to within sight of the vast chain of snowy peaks that divided Bengal from the mysterious lands of Tibet. Rennell named them the Tartarian Moun-

James Rennell by Scott. Illustration from *Major James Rennell and the Rise of Modern English Geography* by Clement R. Markham, 1895.

tains, the name Himalaya then being unfamiliar to westerners. Like all those who first see the Himalayas in the cool season, before the great billowing thunderheads obscure from sight the most spectacular view one can behold, Rennell was awestruck. The stunning peaks of Kangchenjunga, Chomo Yummo, and Chomo Lhari towered higher than the imagination could credit. This all suggested to Rennell that he was seeing peaks that were "among the highest mountains of the old hemisphere."[5]

The mapping methods adopted by Rennell were superior to those of Roy in Scotland and were known, then as now, as route surveys. Instead of surveying a wide tract of country (which was unnecessary for Rennell's purposes), the surveyor built up a detailed map along a narrow strip, such as an access route or waterway, and added the detail of the land on either side of the route by eye or from compass bearings. His tools were the army man's prismatic compass and perambulator. It was this method he employed on the Ganges and completed on his trek back to Calcutta. What made Rennell's maps and methods superior to anything then being produced in Europe was that he applied the navy man's sense for position, establishing latitudes and longitudes by astronomical means for many hundreds of points along his routes. This had the effect of not only accurately locating the surveys but also of controlling the scale, or dimensions, of the work.

On 3 May 1765, Robert Clive, or Lord Clive of Plassey as he now was, returned to Calcutta from London to take command as the governor of Bengal. The tasks before the new peer were colossal, too much for one man and, in the end, would be the death of him. Just one of the many hundreds of things occupying the general's mind and time was the Bengal map project. At Fort William, Clive examined Rennell's work and was much impressed with both the product and the man.

Complying with Rennell's request for additional help with the huge task, Clive issued the necessary instructions. In October 1765, Ensign William Richards of the company's engineers' regiment was appointed as the surveyor-general's assistant. Rennell, an agreeable and affable person himself, was joined by a likewise similar character: "I have now company at all times; and luckily for me, the gentleman proves a very agreeable and cheerful companion."[6] Taking advantage of the break in fieldwork occasioned by the onset of the rains, the surveyor-general and his new assistant set up their headquarters in Dacca, a central location, where the Ganges river system from the west joined the Brahmaputra flowing in from the northeast.

At the start of the 1766 "cold season," Rennell led his survey crew 250 miles north toward Bengal's border with Bhutan and the small province of "Baar" (Cooch Behar, now the western part of Assam). Not far from the forks of the mighty Brahmaputra River, word reached the survey party that one of Rennell's old shipmates from the *America,* Lieutenant Morrison, was leading his troop of 90 sepoys against an armed band of 800 fearsome "Sanashi Fakirs," who were plundering and terrifying the local townships. Rennell, a serving officer in the Indian army, resolved immediately to go to Morrison's aid. When he finally caught up with his navy friend, the lieutenant had already "defeated the Sanashis in a pitched battle." Next morning, after resting under cover of the forest, Morrison, Rennell, and Ensign Richards went off in search of the remaining marauders. Coming across a small village, Morrison decided to reconnoiter. The three Europeans, together with a sepoy adjutant and Rennell's Armenian assistant, cautiously approached the village. Turning a corner into the marketplace, the four men suddenly came face to face with "two lines of the enemy," armed to the teeth and with sabers drawn; the Europeans were quickly surrounded.

Flight was the only prudent recourse. Morrison escaped unhurt; Ensign William Richards fought his way out, receiving a slight wound. The sepoy adjutant fought like a tiger and was badly wounded but survived. The Armenian was less fortunate and was killed outright, with a dagger in his heart. Rennell was cornered by the fakirs and cut off from escape. As the enemy closed in for the kill, he leveled his pistol and fired: it just "flashed in the pan," and now he had nothing but his navy cutlass to defend himself. Slowly, Rennell edged backward, thrusting and slashing with his sword.

It was a desperate fight, and the surveyor received many cuts and wounds from the enemy's sabers. He managed to kill one man, running him through with his sword, and then when he thought only a few of the enemy were at his back, he made a run for it. Morrison, meanwhile, had rounded up his troops and led a charge into the Sanashis "and cut most to pieces." Rennell staggered to the safety of the British line, where he collapsed. He was seriously wounded and had suffered a great loss of blood. His right shoulder blade was cut clean through, with a gash a foot long. His ribs were damaged, and he was cut about the elbow, had a stab wound in the left arm, and suffered a deep cut in the hand that left it permanently incapacitated. Morrison did not have any medical aid with him, and the nearest doctor was at Dacca, 300 miles down the Brahmaputra.

Morrison and Richards wrapped their friend's wounds as best they could, and the soldiers gently laid Rennell in an open boat. They sent the wounded man down the river with a company of rowers and sepoys but with little expectation that they would ever see their friend again. For 6 long, hot days, the Indians guided the boat toward Dacca, applying raw onions to ease his terrible festering wounds. Miraculously, Rennell survived the trip and was delivered, barely alive, into the care of Dr. Francis Russell. For many days, Russell despaired of keeping the surveyor alive, but, bit by bit, Rennell recovered. He slowly regained the use of his arms, and 6 months after the battle, his worst wounds were mostly healed. Although the great loss of blood was a permanent blight on his constitution, he was alive and grateful to Russell for his tireless administrations (they became lifelong friends) and for being "not entirely deprived of the use of my right arm, the provider of my daily bread." When news of Rennell's close brush with death reached the ears of General Clive, the governor immediately issued instructions that a company of sepoys would in future escort all survey parties.

The survey office in Dacca was beginning to fill up with sketch maps, journals, plans, and other geographical documentation. All this information had to be sorted, classified, evaluated, and finally incorporated into the map of Bengal. While William Richards continued with the fieldwork, Rennell occupied his convalescence with preparing the map. Shortly before he departed for England in January 1767, Lord Clive wrote to the directors in London, "We [the Calcutta council] have appointed Captain Rennell, a young man of distinguished merit in this branch, to be Surveyor General, and directed him to form one general chart from those already made. This though attended with great labour does not prevent him from prosecuting his own surveys, the fatigue of which with the desperate wounds he has lately received in one of them, have already left him but a shattered constitution."[7] From this date, Major James Rennell was officially known as the surveyor-general of Bengal.

The 1767–1768 season saw a much recovered Rennell working his way through the country east of the Brahmaputra, through Rangpur and Rangamati, and along the right bank of the Bhagirathi River. Once more a violent encounter seemed unavoidable when Rennell stumbled into a large detachment of the Bhutan army, ready to deny the English surveyors access to their lands. Narrowly avoiding one ambush, Rennell and his men then fell into another. In the skirmish with the Bhutanese warriors, one of his sepoys was seriously injured. The survey party retreated into the dense forests to avoid further clashes and made their way out of the disputed region.

The next season Rennell took the surveyors even further eastward along the valley of the Brahmaputra into a wild country "infested by savage animals."[8] This was the remote Duars (the Doors) of the Assam valley, notorious for its tiger-haunted forests and wild elephants, monkeys, scorpions, and snakes. Rennell's route survey followed a narrow tongue of land tucked between the foothills of the Himalayas on the north and the highlands of Meghalaya on the south. As the survey party worked its way along the edge of a belt of forest, a leopard suddenly appeared from out of the trees. The big cat had silently stalked the party before it attacked. The leopard slashed and bit its way through the troop, seriously injuring several of the men before it turned its attention to Rennell. Leopards are not as large as a Bengal tiger, but on the other hand, they are not small, weighing in at some 200 pounds, and move with great speed and terrifying agility. As the leopard sprang, Rennell, with extraordinary presence of mind, thrust his bayonet down the animal's throat.

By the close of the 1770 season, Rennell had been working across the northeast of India for some 6 years. He was only 28 years old, but the encounter with the Sanashis had left him with a shattered constitution and weakened body. He became increasingly prone to attacks of different kinds—from ague and tropical fevers, mostly malaria. As a consequence of recurring illness, he spent less time in the field and the burden of surveying fell increasingly on the shoulders of William Richards.

In the same year a terrible famine raged across Bengal that lasted almost 2 years; 15 percent, some estimate 20 percent, of the native population perished from disease and hunger. Europeans were no less immune to the ravages of disease, and as the death toll rose, so the revenues of the East India Company fell. In contrast to its declining income, its running costs, especially the costs of its

Surveying through the tiger-haunted forests
of northeastern India.

army, spiraled out of control. This disaster coincided with a time of difficulty, economic stagnation, and trade collapse across Europe.

The company was in crisis, bordering on the very brink of ruin. To stave off a collapse, with all the consequences that would follow, the London Court of Directors appealed for help from the British government. The government's first response was the 1773 Tea Act, which resulted in the Boston Tea Party, which in turn fueled the flames of the American War of Independence and Britain's loss of its first colonies; the circle was closing. The 1773 Regulating Act for India followed, marking the British government's first step toward replacing the company's power base in India with direct British rule. Although the government monitored the company's activities and general policy, it did not assume any degree of power. The governor-general and council maintained control over the territories of Madras, Bombay, and Bengal. The overall management of the company rested, as ever, with the so-called Court of Directors, but now Parliament insisted on knowing everything of import that was of a civil or military nature. Many of the ills that had besotted the honorable company were corrected, including the justice system and the rampant bribery and corruption among the company's servants.

During the period of famine, Rennell worked on the map of Bengal from his house in Dacca, interspersed with the occasional trip to his office at the fort in Calcutta. He was a frequent visitor at the home of Governor Cartier, and it was there that he made the acquaintance of Jane Thackeray, aunt to the future novelist William Makepeace Thackeray. After a year's engagement, they were married and made their home in Dacca. During the hottest part of the year, when Dacca became particularly unhealthy, they retired to a house they had acquired in the cooler highlands of Islamabad, not too distant from the sea.

Under the terms of the 1773 Regulating Act, the East India Company's first "new wave" governor-general, Warren Hastings, was appointed. Hastings was a man who understood India and its people better than did any Briton of his day. When governor of Bengal, he had established a sort of civil service, employing British tax collectors to replace the "corrupt" practices of native collectors. Fortunately, Hastings's new broom did not brush too harshly across the surveyor-general's floors, and it seems that relationships between the two men were extremely cordial.

Rennell now spent all his time preparing maps for printing and no longer ventured abroad into the wild. His health was precarious, and it must have been obvious to all who knew him that another season in Bengal would mean his death. By the close of 1776, Rennell was satisfied that his work in India was done and that he had sufficient information for his map to return home. Warren Hastings, who had come to value Rennell's work very highly, bestowed on him a pension of £600 a year.

The Rennells left India in March 1777, arriving at the company's waystation of St Helena via the Cape of Good Hope in late September. Jane Rennell was heavily pregnant, so it was decided they would remain at the governor's house on the inclement Atlantic island until the child was born. In mid-December, with their tiny baby girl, the couple set off on the final leg of their long journey to England. It was not without incident: narrowly averting shipwreck, their East Indiaman was almost overwhelmed during a violent storm. On 12 February 1778, nearly a whole year after leaving the blistering, torrid heat of Bengal, they arrived to a freezing cold, wet, and windy British winter.

The onset of the winter of 1778 marked a reversal in American fortunes in its War for Independence. In May 1779 came news that the British had burned Portsmouth and Norfolk, Virginia. Then it was Britain's turn for bad news when its other old adversary, Spain, declared war as an ally of France but, bringing some relief, had not chosen to align with the Americans. Another major defeat for Washington's army occurred in September, when the Americans tried to re-cover Savannah.

With the coming of spring 1780, the British opened the campaign with a seaboard assault and captured the city of Charleston. Washington countered by sending a large force of reinforcements, only to suffer his worst defeat of the war—losing his entire army in the south. Then, in another reversal of fortune, the Americans defeated the British at the battle of Springfield, New Jersey. In July the French fleet, commanded by the veteran Count de Rochambeau, landed 6,000 fresh troops and artillery at Newport but were almost immedi-ately blockaded by the British fleet.

Prime Minister Lord North's problems at home were almost as bad as they were overseas. His ministry was plagued with factional division. Among the more notable domestic incidents of 1780 were the Gordon Riots, so called after Lord George Gordon, an ex-naval officer, religious bigot, and leading opponent of Catholic emancipation, led a crowd of 50,000 to present a petition to Parlia-ment for the abolition of the 1778 Roman Catholic Relief Act. Five days of fierce rioting resulted, and considerable damage was done to the property of Catholics, to the Bank of England, and to many of London's principal prisons.

In the new year of 1781, Cornwallis and his 10,000 exhausted troops arrived at Yorktown on Chesapeake Bay. George Washington suddenly saw his oppor-tunity, and in a sudden change of strategy made straight for Yorktown. On 30 August, the French fleet arrived and de Grasse joined up with General Lafayette's American troops. Cornwallis was trapped; the British naval squadron under Ad-miral Graves was in disarray. Suddenly it was all over, a "defeat snatched from the jaws of victory," but a most unlikely victory. However, as William Pitt had foreseen, victory over America was an impossible dream.

James Rennell learned of the defeat at his new home in Charles Street, off Cavendish Square. By then he was firmly established within London's circle of the intelligentsia, an academic at heart. He was not particularly politically minded but was very much opposed to the continuance of the American war and the hardships it was causing to ordinary people: "I hear of nothing but mis-ery and want among the lower orders, and yet we are said to be in a flourishing condition. To hear my Lord North declare it, after exhausting his country, is too much for my patience."[9]

When the news of Cornwallis's defeat reached the prime minister, North wrote in his diary, "Oh God! It's all over." By the end of February 1782, Parlia-ment had voted to curtail the unpopular war. On 20 March, Lord North re-signed and Lord Rockingham and his Whigs took power. The peace talks were held in Paris, with Benjamin Franklin representing the new United States of

America. In January 1783, Britain signed a preliminary peace treaty with France and Spain, and on 3 September the Treaty of Paris was signed by the United States and Great Britain.

Many scale maps of North America were available before the Revolutionary War, and many more were made during the campaigns for the use of the generals at home and in the field. Maps of the campaigns, marches, and counter-marches proliferated, yet, for all this, the information upon which they were based was, for the most part, without any scientific credibility. The American war had demonstrated again the problems that generals faced when moving their "modern armies" about poorly mapped foreign country. In the first years of the war, the protagonists on both sides were hampered by a lack of good, reliable, and up-to-date information. George Washington was acutely aware of the fact and did something about it. In August 1781, he engaged a Dutch émigré land surveyor, Simeon DeWitt, as surveyor-general to the Continental Army.[10] After the war, DeWitt's maps would be added to the growing library of American geographical knowledge and used to augment the road maps of Christopher Colles, the Irish-born mathematician and surveyor who did so much to improve America's early road maps.

The inadequacy and availability of reliable military maps for the American campaign was not lost on the returning British generals. France had excellent maps. Peter the Great, tsar of all the Russias, had the French cartographers Joseph Nicolas and Louis Delisle. Sweden's map was well advanced, as was its Scandinavian neighbor Denmark. Even Norway, the impoverished vassal of Denmark, had begun a project, as had the Austro-Hungarian Empire, the powerhouse of Europe and tinderbox of European politics. But Britain, in stark contrast to its assiduous European neighbors, many of whom were potential foes, did nothing.

British army men returning home on leave from India were lauding their fine maps and charts, praising the splendid work of Rennell, Richards, and the many other surveyors, hydrographers, and geodesists that were yearly transposing a vast continent onto flat sheets of print. When, people were asking, would the woolly-minded British government wake up to the fact that they were in the eighteenth century?

Rennell's departure for England was by no means the end of surveying in India. Having seen the benefits of good maps and charts, the East India Company was desperate for even better information. They needed to complete the surveys of the territories they already controlled and to learn about the geography of the entire subcontinent. Their ships' captains were demanding better charts in order to avoid the dangers of the treacherous coasts.[11] Another war had also broken out, this time against the Marathas, and once again commanders wailed about the absence of topographic information. Colonel Thomas Goddard, leading a forced march to relieve the army that was beleaguered at Surat, was obliged to cover 300 miles (500 km) in 19 days "through regions unknown in England and untraced on our maps."[12]

# 18

## LATE BY A WHOLE YEAR

PRESENTING MAJOR JAMES RENNELL with the Royal Society's Copley Medal for 1791, the president of the society, Sir Joseph Banks, said:

> I should rejoice could I say that Britons, fond as they are of being considered by surrounding nations as taking the lead in scientific improvements, could boast a general map of their island as well executed as Major Rennell's delineation of Bengal and Bahar: a tract of country considerably larger in extent than the whole of Great Britain and Ireland; but it would be injustice to the Major's industry were I not here to state that the districts he has perambulated and planned exceed, probably, in extent the whole tract of surveyed country to be found in the maps of the European kingdoms put together, while the accuracy of his particular surveys stands yet unrivalled by the most laborious performance of the best county maps this nation has hitherto been able to produce.[1]

Banks's accolade for Rennell's wonderful work was well deserved, but it was also a political statement aimed at the British government in support of his old friend William Roy's aspirations for a national map of the British Isles. To some extent, Banks was also being unfair to the country's private mapmakers who were already producing good material of the counties and shires. Indeed, Roy never envisaged a government-funded mapping organization per se. Rather as the railways supported trade through infrastructure, his vision was for developing a framework of rigid triangulation on which private county and town mapmakers could base their work.

The president of the Royal Society's claim that Britain could boast a "lead in scientific improvements" was a reference to Roy's great trigonometrical survey. This, the first British trigonometrical scheme, was not the result of the British government's foresight, but rather a nationalistic reaction to a memoir sent from France.

On 7 October 1783, a month after signing the peace, Count d'Adhémar, the French ambassador to the court of St. James, delivered a *mémoire* to Secretary of State Charles Fox, addressed to King George, from César-François Cassini de Thury, directeur de l'Observatoire Royale de Paris: *Il est intéressant pour le progrès de l'astronomie que l'on connaisse exactement la différence de longitude et de latitude entre les deux plus fameux observatoires de l'Europe; et quoique les observations astronomiques faites depuis un siècle offrent un moyen assez exact pour parvenir à cette recherche, il parait cependant que l'on n'est point d'accord sur la longitude de Greenwich à onze seconds près, et sur la latitude à quinze seconds.*[2]

The objective was clear: to tie together the observatories at Paris and London so there was no doubt about their true positions on the earth. The latitude of each observatory had been established by star observations, but as everyone now understood, latitude could be corrupted by gravity anomalies. The longitude of the observatories in both cases was, of course, $0°0'0''$. Only by measuring the difference in time between the two could a relative longitude separation be achieved. But Cassini had a grander plan in mind—to physically measure the difference in both latitude and longitude through a great scheme of triangles to further improve knowledge on the size and shape of the earth.

The opening paragraph of the *mémoire,* suggesting that the latitude of the Royal Observatory at Greenwich was in doubt by some 15 seconds of arc and that its correct position could be derived only by a direct connection with the Paris Observatory, was somewhat tactless and incorrect. Under other circumstances, it might have led to the project's rejection out of hand. As it was, when Fox showed the note to Sir Joseph Banks, the president of the Royal Society immediately saw the opportunity it proffered.

The note, written by Cassini some time before the cessation of Anglo-French hostilities, described how the academicians of France had extended their triangulation scheme from Paris all the way to Calais on the French coast. They had also measured the distance from Calais to Dover Castle by observing two large triangles, the difference of which was just two toises. This must have made disturbing reading for Fox, learning that the French had such advanced technology. Cassini appealed to the British sense of pride by noting the country's good fortune in having *d'un Souverain qui aime les sciences* and boasting great explorers such as the *célèbre Cook.* He need not have bothered: the war was at an end. King George III was desperate to build bridges with the French monarch and could be relied on to give the project his full support.

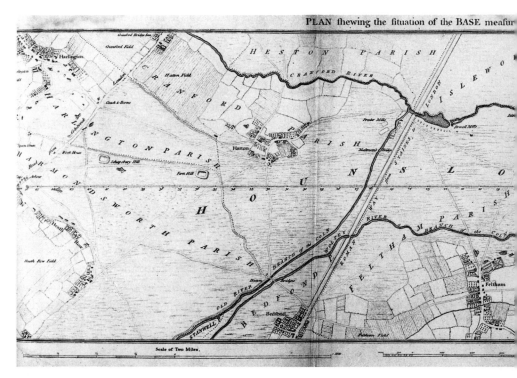

Early ordnance map showing the Hounslow baseline illustrated in *Philosophical Transactions of the Royal Society* 75. Courtesy Ordnance Survey. © Crown Copyright.

In London that fall of 1783, William Roy was, for his "own private amusement," measuring a baseline across the fields between the Jew's Harp, near Marylebone, to Black Lane, close by Saint Pancras. When he heard about the French *mémoire,* he immediately realized that here was the excuse he needed for "laying the foundation of a general survey of the British Islands."[3] Within weeks, the Royal Society had formed a small committee to discuss the French proposal. It was agreed unanimously that Colonel Roy (he was promoted in 1777) should undertake the business. A great deal of new equipment was needed, and King George, a keen amateur scientist, agreed magnanimously to fund the manufacture of the necessary instruments.

So that Roy could be released from his military duties to undertake the work, it was necessary for him to secure the permission of his superior, the duke of Richmond, master-general of the Board of Ordnance. Charles Lennox, third duke of Richmond, can rightly be accorded the honor of being the true founder of Britain's national survey. Born in 1735, he had inherited the family title in

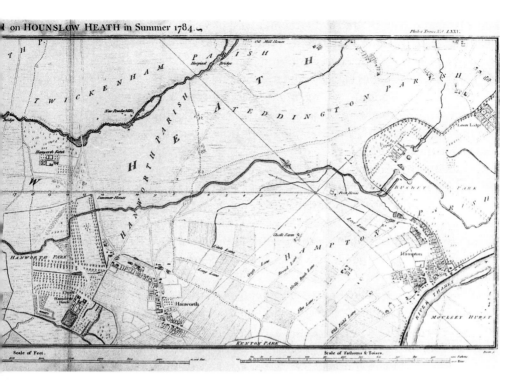

1750, shortly before he entered the army at the tender age of 15. Like Roy, he was at the battle of Minden, where he distinguished himself, and after the Peace of 1763, had gone to Paris as Britain's ambassador. He was an outspoken critic of Britain's war with America, earning the displeasure of the king. In the course of time the rift was mended, and they became good friends. A great supporter of all of Roy's trigonometrical ambitions, the duke was also a fellow of the Royal Society and immediately endorsed Roy's nomination for undertaking the survey.

The plan called for a chain of triangles to link the Royal Observatory in Greenwich Park to points on the southeast coast of England, from where triangles could be formed that would connect across the English Channel to France. The first task, therefore, was to establish a baseline from which the triangles could be formed; Roy selected Hounslow Heath, 20 miles west of London. The heath, once a barren waste and haunt of highwaymen, had been mostly "enclosed" to form a rolling landscape of regular fields and smallholdings. It was well to the west of the metropolis and conveniently situated on the road to Bath. On the morning of 16 April 1784, a group of distinguished gentlemen assembled at King's Arbour, an orchard just behind the Three Magpies public house (which still exists) and a quarter mile west of Sipson Green. Nearby was Heathrow Hall, the farm that gave its rustic name to the sprawl of London's Heathrow Airport. Joining Roy that morning was Sir Joseph Banks, president of the Royal Society, Charles Blagden, and the amazingly shy but brilliant mathematician Henry Cavendish. Setting off in jovial mood, the men strolled across the fields of the heath toward the southeast, inspecting the ground as they went. Five miles later they arrived at Hampton Poor House, near the gates to Bushy Park beside the River Thames.

Satisfied with the route, the next step was to clear and prepare the ground for the measurement. Instead of employing local laborers, it was decided on the grounds of security and economy to employ soldiers. This had the benefit of adding a military air to what was otherwise a civilian undertaking. Accordingly, on 26 May a party of ten men of the Twelfth Regiment of Foot, together with their sergeant and a corporal, marched from nearby Windsor Castle to set up camp on the heath.

While the soldiers cleared and leveled the ground, a pair of 20-foot-long deal rods were fashioned from "the best seasoned timber, from an old mast cut up on purpose."[4] Just as Mason, Dixon, Maskelyne, and Burrow had discovered before when using fir rods for measuring baselines, humidity affected the length of the wood; this unfortunate consequence caused Roy great concern. Fortuitously, one of the volunteers for the measurement, Lieutenant-Colonel Calderwood of the Horse Guards, had a stroke of genius. Instead of fir rods, Calderwood suggested using glass tubes, the length of which would not be affected by dampness and only required adjusting for temperature variations. Immediately, Calderwood was "requested to make a trial at the glasshouse, as soon as possible after his return to town." The very next day, the glassmakers at the filthy, noxious factory, located in what is today fashionable Mayfair, succeeded in drawing an 18-foot-long tube, an inch in diameter. Several of these glass tubes were made and sent to the instrument maker Jesse Ramsden for finishing off.[5]

Meanwhile, the baseline was measured with a new chain made especially for the survey by Ramsden (he was doing well from the survey). The measure was only approximate and was taken across the ground, starting on 16 June and completed 6 days later. Ramsden, with an unusual show of alacrity, delivered the finished glass tubes, ready for the precise measurement to begin. On 17 August, the tubes were carefully laid within wooden troughs supported on stands and were calibrated against the 42-inch brass standard that had been made by John Bird 30 years before. This was the same imperial standard that Bird used as a template for the standard used by Mason and Dixon in America and by Maskelyne on Schiehallion. By this simple coincidence, if coincidence it was, the standard foot used in America, on Schiehallion, in northern India, and now, at the start of the great trigonometrical survey of Britain itself, was identical.

The baseline measurement caused much excitement in scientific circles and attracted numerous visitors, including King George, who turned up on 21 August to inspect progress. Joseph Banks was a frequent visitor, providing tents for the guests and laying on "elegant refreshment" and entertainment. On 30 August, the baseline measurement was at last finished. The resulting distance as measured with the glass rods, corrected for temperature, was calculated by Roy to be 27,404.08 feet (8,353 m). The terminal points of the baseline were marked with wooden tubes sunk into the ground, replaced some years later with iron cannons from Woolwich Arsenal.

At about this time, or perhaps a little earlier, William Roy was seeking a mathematical assistant. One story has it that it was Jesse Ramsden, the instrument maker, who first recommended Isaac Dalby. This probably took place in 1780 or 1781 since Dalby was then teaching at the Royal Naval School in Chelsea and probably Roy knew him or was acquainted with his skills. The astronomical and

Canon that marked the King's Arbour terminal of the Hounslow Heath baseline, now restored to its original location at Heathrow Airport.
Courtesy Ordnance Survey. © Crown Copyright.

mathematical community in London was intimate and well connected. However it came about, what is certain is that in 1784,[6] Dalby joined Roy as his full-time assistant and official mathematician of the survey. Shortly before his death, Roy wrote that "it is proper that I should mention that Mr Dalby, who had been recommended as an assistant, has acquitted himself throughout the whole perfectly to my satisfaction."[7]

Isaac Dalby (1744–1824), like Charles Mason, was a Gloucestershire lad made good. He had arrived in London in 1772 and first found employment as an arithmetic teacher at Archbishop Tenison's grammar school in Charing Cross before accepting the astronomical observer's position in Lord Topham Beauclerk's observatory. Beauclerk—a friend of Samuel Johnson, a rake, immensely wealthy, and with immaculate connections—was a keen amateur polymath and had one of the greatest libraries in the land. The combination of eloquent employer, abundant academic materials, and an endless stream of interesting visitors was a mixture that provided Dalby with an interesting and stimulating career and brought him into contact with many useful people. In 1780 Beauclerk died, and the following year Dalby took up the position of mathematical master at the Royal Naval School at Chelsea.

In August 1784, the month in which Roy completed the Hounslow Heath's baseline, he requested the Royal Society to commission Jesse Ramsden to construct "a large circular instrument" for measuring the angles of the proposed network of triangles. Ramsden, the son of a Halifax innkeeper, was a perfectionist and, in 1777, had developed a mechanical means of dividing circles with great accuracy, thus removing the small errors that had plagued hand-divided circles. For this scientific breakthrough he was awarded the Royal Society's Copely Medal.

A kindly and generous man, he also had the "artist's genius distaining time restrictions," a characteristic that was to sorely try Roy's patience. It is said that, on one occasion, Ramsden was invited by the king to an audience at Buckingham House. He arrived "precisely as he supposed at the time named in the royal mandate. The King remarked that he was punctual as to the day and hour, while late by a whole year."[8] Jesse Ramsden's name was almost legendary, and his reputation for making fine instruments and mechanical models had brought him into an influential circle of "practical philosophers" whose talents, skills, and dogged determination (and often reckless disregard of costs) were rapidly transforming Britain. At the west end of the Strand, where most of the scientific manufacturers had their businesses, was St. Martin's Lane, with its fashionable coffeehouses. There, at Slaughter's Coffee House, Ramsden met with some of the most ingenious of the new "engineers," men such as Matthew Boulton of Birmingham and his partner James Watt, who were revolutionizing manufacturing with their mighty steam engines. His circle of friends included John Smeaton, the great civil engineer for whom Ramsden made working models, Joseph Banks, and the potter Josiah Wedgwood.

On 24 August 1784, Jesse Ramsden wrote to one of his clients, John Goodriche, explaining why delivery of the gentleman's "Equatoreal instrument" was being delayed. He was, said Ramsden, working almost exclusively on the Royal Society's instrument "for ascertaining Geometrically the distance between Paris & London." He gave an interesting account of the work and the difficulties they were experiencing:

> . . . This being done at the King's expence no care or pain . . . has been spar'd to do everything in the best manner and indeed I have been oblig'd to make the [glass] Rods twice over. Everyone here were convinced that Straight grain'd Fir perfectly free from knots wou'd not expand Longitudinally but after having made all the Rods in the best manner and so as to prevent them from bending and the Fir being the best the Kings Dock Yards could afford yet the expansion from the moisture was too great to permit us to use them for this mensuration & have since constructed them anew & made them with glass tubes of 20 feet Long each.[9]

While the great theodolite was being constructed, Roy and his small team used the time to reconnoiter the land between London and the south coast, selecting suitable sites for observing stations and locating and surveying a piece of land on Romney Marsh preparatory to measuring a base of verification. The reconnaissance was a crucial part of the survey planning that Roy had learned from his military training and during the survey of Scotland.

The reconnaissance was also useful to Roy in his capacity as a senior Board of Ordnance officer, allowing him to visit and inspect some of the military strongholds and fortifications of southern England. At the very front line of Britain's defense was the port of Dover, the nearest point to France and so considerably positioned to a nicety by the French geodesists. Visiting the great Norman castle high above the town in January 1785, he would have beheld a very peculiar sight—a glimpse of the future and how, one day, mapmakers would escape the confines of earth to make their measurements from the air.

There had been reports from France that two brothers, Joseph and Jacques Montgolfier, from Annonay in the Rhône Valley, had been experimenting with aerial balloons. In June 1783, they had filled a "giant Globe" of taffeta and paper with "lifting smoke" and had risen over a vertical mile through the atmosphere.

The brother's claims were verified when, 3 months later, they sent aloft a duck, rooster, and sheep in front of their doomed majesties King Louis XVI and Marie-Antoinette. In November came the staggering news that two men, Pilatre de Rozier and the Marquis d'Arlandes, had voluntarily stepped into the gallery of one of Montgolfier's cotton and paper balloons, inflated to bloated proportions by the application of heat and smoke from a brazier. When the balloon was released, the men became the first human beings to step voluntarily off their planet and into the air. Benjamin Franklin, then in Paris, sent graphic reports of the experiments to his friend Joseph Banks at the Royal Society.

Suddenly, balloon mania gripped both nations, and within months the hydrogen gas balloon was competing with the primitive hot-air balloons. In Derby, Erasmus Darwin, Charles Darwin's grandfather, built and was the first Englishman to fly a hydrogen balloon. Not to be outdone by his French "cousins," in September 1784 the prince of Wales and a huge crowd gathered at Chelsea to watch Vincenzo Lunardi lift off on a flight that took him all the way to Hertfordshire. In December, James Watt wrote to James Lind with the results of an experiment conducted by his partner, Matthew Boulton, with an exploding balloon, supposedly to test whether the result emulated thunder—apparently, it did.

On 7 January 1785, at Dover Castle, Jean Pierre Blanchard (1753–1809), an egotistical, French émigré inventor (his claims to fame included *le velocipede,* an early form of bicycle, and the *visseau volant,* an impossible "flying vessel"), was preparing one of the new gas balloons preparatory to being the first person to fly across the English Channel. His preparations were going ahead within a barricaded compound constructed to keep his sponsor, Dr. John Jeffries, an American businessman living in England, from joining him.

Jeffries was as determined as Blanchard to make the flight and hired some sailors from the coastal town to break into the compound. Fortunately, it didn't come to violence and an amicable agreement was made: if the balloon was too heavy, Jeffries agreed he would jump into the sea. But Blanchard had one last trick up his sleeve. As the balloon was readied for flight, he announced that it was overweight with Jeffries on board. The doctor, familiar with Blanchard's conniving ways and referring to him as "a petulant little fellow . . . physically well suited for vapourish regions," searched the aeronaut and discovered the problem—Blanchard was wearing a belt of lead. Freed of the excess weight, the gas balloon rose from the ground and began its hazardous journey, sailing to the southeast. The crowds gathered on the cliff top cheered and clapped; the last they saw of the giant balloon was as a small blob far away, sinking slowly toward the water.

In fact, Blanchard and Jeffries did make it and were the first men to fly across the English Channel; they had to jettison all their ballast, baggage, and cargo to lighten the balloon. Blanchard even discarded his clothes. In this last matter, Jeffries refused to comply, saying he would "rather drown than face a Frenchman in the nude." The rapid approach of the sea's surface changed his mind and his clothes soon joined those of Blanchard, floating away on the current. Jeffries then had an inspiration for further reducing weight by releasing their bodily fluids into the English Channel.

Among his many experiments and trials that winter preparatory to executing the trigonometrical survey, Roy was experimenting with "white lights," a volatile composition of "28 parts Nitre, 4 parts of Sulphur and two parts of orpiment powdered fine,"[10] which, when fired in a copper vessel, produced a brilliant white light. Roy proposed to use these lights at night as observing targets for his new theodolite. He fired one at Dover Castle to test the results. By this date, James Lind, Maskelyne and Roy's friend from Edinburgh, was already living at Windsor, where he was the physician to the royal household. Sadly, the fat, jolly Mrs. Lind had passed away. Naturally, Roy recruited his old friend as a member of his team of volunteers and used him, and his personal equipment, as an observer of the white lights and for making sundry measurements.

In April 1785, Joseph Banks finally got around to showing Nevil Maskelyne the *mémoire* from Cassini (who by that time was dead) and invited the astronomer royal to prepare his comments for publication. Banks's tardiness in letting Maskelyne see the note was, ostensibly, the consequence of the unpleasant disagreement within the Royal Society. In 1778, Banks had commanded a sufficient majority of the fellows to be elected president. Almost immediately, the society polarized into the "Men of Science versus Macaronis."[11] Sparked by Banks and supported by Charles Blagden, the "macaronis" had accused Maskelyne's mathematical friend Charles Hutton, the society's foreign secretary and someone Banks greatly disliked, of neglecting his duty.

The acrimonious dispute dragged on into early 1785 and resulted in the principled resignation of Hutton and other distinguished academics. The immensely shy and retiring Henry Cavendish was so affronted that he proposed a vote of confidence in Banks. When an opportunity arose to elect a new secretary, the mathematicians' man was Charles Hutton and the macaronis' candidate was Charles Blagden; Hutton lost.

Even before Banks was elected president of the Royal Society, he was carving out a position of influence in society matters and in politics in general. He was actively involved in almost every aspect of Pacific exploration and in the beginnings of the colonizing of Australia. In 1778, he lent his support to the creation of a settlement for miscreants at Botany Bay, a place that grew to become the notorious penal colony. He was not always the best judge of character, proposing Cook's sailing master, William Bligh, for the expeditions to transport breadfruit to the slave plantations of the West Indies that ended in the ignominious mutiny aboard HMS *Bounty* in April 1789 and the subsequent loss of the ship.

Banks's other possible reason for the delay in showing Dr. Maskelyne (Maskelyne was awarded the degree of doctor of divinity in 1777) Cassini's note was that he knew the astronomer royal would object strongly to the notion that the geographical position of his beloved observatory was in error. Consequently, reasoned Banks, Maskelyne might, in a fit of pique, have written a paper denouncing Cassini's rash pronouncement and thereby jeopardizing the Royal Society's justification for supporting the great triangulation campaign. No doubt the astronomer royal's first comments when he read the *mémoire* were unrepeatable. However, once he had composed himself, he considered the matter in depth. On the question of the latitude, he had no doubts that Cassini was mistaken, and Maskelyne put the difference at "no worse than 4."5";[12] but he was not so confident about the longitude.

Between 20 September and 3 October 1785, Maskelyne sent his observatory assistant Joseph Lindley on a clandestine "chronometer run" to the Paris observatory.[13] Four chronometers and four precision watches, all made by John Arnold, were calibrated against the transit clock at the Greenwich Observatory before being boxed up for Lindley's trip to France. Traveling by coach and channel packet, it took him 6 days to reach Paris. There the "watches" were surreptitiously compared with the observatory's transit clock. Six days later he was back to London, presenting the results to Maskelyne. From a comparison of the results, the astronomer royal was confident that his difference of longitude on Paris was 9 minutes 20 seconds in time, or in angular terms 2°20′, and would be "within a very few seconds of the truth."

On 10 September 1786, the first signs appeared of Roy's irritation with instrument maker Jesse Ramsden over the time it was taking to build the theodolite. In a letter to James Lind, he wrote, "It will be yet some days before Ramsden can possibly finish the Division and after that the semicircle for the uppermost telescope is to divide . . . and it will certainly render it too late in the season to think of taking men into the Field to encamp. It is hard upon me to have this operation hanging over my head for another year, without any fault my own; But with such a man as Ramsden there is no help for it."[14] Six months later, he wrote again to Lind, complaining about Ramsden's tardiness. In a temper close

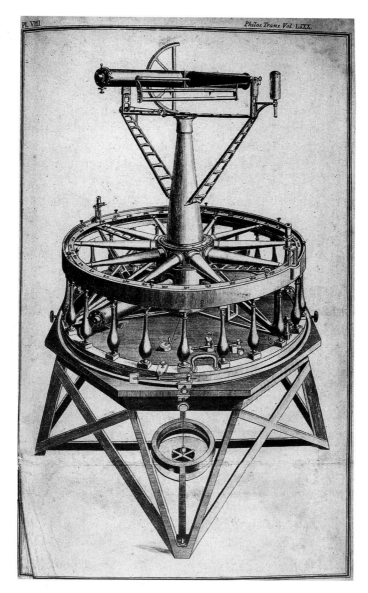

Ramsden's 36-inch theodolite built for the Royal Society.
Illustration from *Philosophical Transactions of the Royal Society* 80.
Courtesy Ordnance Survey. © Crown Copyright.

to the breaking point, Roy visited Ramsden and read the riot act, demanding that the instrument maker rope in "every hand that he can employ" to finish the theodolite.

On 2 July 1787, 3 years from the time the order went in for the instrument, Roy wrote to his friend with some hope: "Although, when one has, unfortunately, to do with such a man as Ramsden, it is altogether impossible to answer for what may happen; yet I hope we shall be able to try the Instrument in Hyde Park on Wednesday or Thursday next." Lind was then instructed to erect a signal staff on the Wardrobe Tower of Windsor Castle for the first experimental observation.

To be fair to Ramsden, the instrument he was building was unprecedented in both size and accuracy. The man was a perfectionist, and the final product would turn out to be well worth the wait, marking, as it did, a turning point in earth-measuring technology. Everything seemed to be on course. Joseph Banks wrote to the Académie Royale, advising them of progress in Britain and asking them to be ready to send across their geodesists. The master-general of the ordnance gave instructions to the arsenal at Woolwich to "supply whatever fireworks might be wanted for signals" and provided General Roy with a detachment of artillerymen to assist with the laboring and to guard the instruments, thus further associating precision surveying with the military arts.

The theodolite was delivered to Roy and successfully tested in Hyde Park. The instrument had a brass circle divided into 10-minute intervals and, by means of micrometers and strong magnifying glasses, could be read to one-tenth of a second of arc. It also was fitted with a powerful telescope, magnifying seventy times, with a fancy new Dolland achromatic lens that enabled far-distant targets to be seen with great clarity and precision. At 200 pounds, the instrument was no lightweight, but Ramsden had "united in it the powers of a theodolite, a quadrant, and a transit instrument."

The instrument used by the French academicians was quite different, although no less accurate within itself, employing what was called a Borda circle, or "circle of repetition," that mechanically added together repeated observations to a target in a way that "could diminish the error arising from the division of the instrument [the French still hand-divided their circles] to an infinite degree."[15] Invented by the French physicist Jean-Charles de Borda, the instrument was much smaller and lighter than Ramsden's giant, but its two telescopes suffered from low magnification, whereas the telescope of Roy's theodolite was powerful enough to easily resolve targets at a distance of 30 miles (50 km).

De Borda's repeating circle, while a clever piece of equipment, relied on repetition for accuracy rather than the absolute precision Ramsden's instrument en-

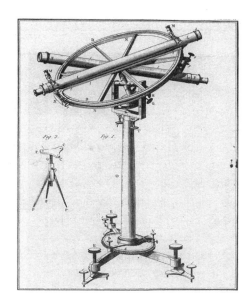

De Borda's repeating circle instrument. Illustration from Cassini, Legendre, and Méchain's *Exposé des Opérations faites en France,* Paris, 1791.

joyed. The Borda circle continued to be refined and was employed by French academicians in their extraordinary scheme to derive a new and natural unit of measurement, the meter.

On 31 July 1787, Major-General William Roy (he was promoted in 1786), assisted by Isaac Dalby and together with a number of the Royal Society's fellows, gathered near Hampton Poor House to set up the great theodolite (for, indeed, it was the grandfather of all modern theodolites) and begin the first national triangulation of Great Britain. The instrument was used to observe church steeples and the tall flagstaffs carried about and set up by the squad of artillerymen. At night or in hazy weather, the brilliant white lights were used. So successful were these that Jacques-Dominique Cassini claimed to have seen one shining brightly in poor weather across the English Channel, 40 miles distant.

It was suspected from the outset that Ramsden's theodolite was sufficiently accurate to detect the very curvature of the earth, and Roy therefore insisted that the utmost care be taken to ensure that the instrument was located exactly above the spot previously occupied by a target. His precautions were well founded for, by the time Dalby computed the sum of the angles in the larger triangles, there could be no doubt; the sum of the angles always exceeded 180° by a few seconds of arc.

Never had there been any instrument that could directly detect the curvature of the earth's surface. The explanation for this phenomenon, known as the "spherical excess," is that the sum of the interior angles of a spherical triangle, a figure measured on the earth's curved surface, is always greater than the constant 180° of the plane triangle. A simple mind experiment demonstrates the grossness of the principle.

An observer standing at the point where the Greenwich meridian (0° longitude) intersects the equator (0° latitude) would observe an angle of 90° between the equator and north. If he or she then traveled due west along the equator to, say, 90° west, the longitude that passes roughly through Texas, the angle between where they had just come and north would also be 90°. Then, to close the triangle, if they traveled to the North Pole and observed the angle between 0° longitude and 90° longitude (the two points on the equator), the angle would again be 90°; 90° multiplied by three equals 270°, or in other words, the spherical excess of the triangle would be 90°. Now, to put things in perspective, it requires a triangle with an area of about 75 square miles (200 km²) to produce a spherical excess of just 1 second of arc.

It must have been very satisfying to Roy, and not least to Jesse Ramsden, that the theodolite was so incredibly accurate. Because the instrument was capable of detecting the spherical excess, Roy developed a theorem for adjusting his triangles to make them "plane," that is, to make their angles add up to 180°, so that they could be solved by normal geometry. Compare the accuracy of Ramsden's new theodolite with that of the instruments he constructed for the Schiehallion experiment 14 years before. With those instruments, Burrow was fortunate to close a triangle to within a couple of minutes of arc; the great theodolite was some 600 times more accurate.

The chain of triangles Roy and Dalby designed had apexes across southeast England, extending from Hounslow Heath across Surrey and Kent to the coast and the Romney Marsh baseline. The all-important connection to the Royal Observatory was made in August 1787. A temporary scaffold was erected above the transit instrument, housed in a building on the southeast side of the observatory, on which the theodolite was mounted. Between 14 August and 3 September, Maskelyne was host to Roy, his military assistants, and the gentlemen who were helping him with the observations, whose number included Henry Cavendish, Isaac Dalby, and Maskelyne's bête noir, Charles Blagden.

In their correspondence with the French academicians of the Paris Observatory, the Royal Society had set the date of 20 September for the historic rendezvous at Dover. Leaving Dalby in charge of the observation team at Wrotham Hill in Kent, Roy and Blagden traveled on the "Flying Machine," the stagecoach that ran daily between London and Dover. Stormy weather had prevented the French contingent from crossing the channel, and it wasn't until 23 September, the day

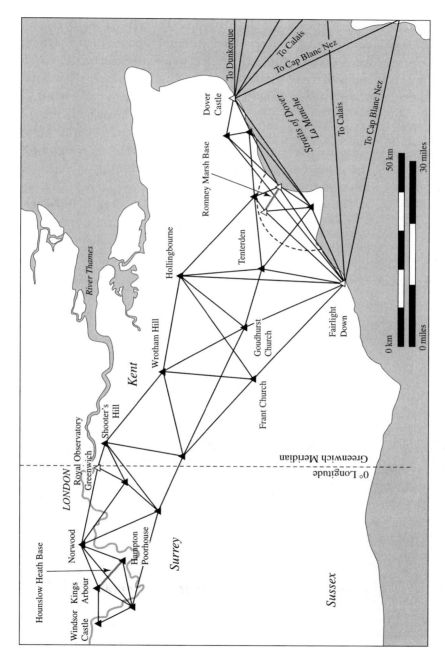

Roy's trigonometrical survey of southeastern England.

before the town's annual Mop Fair, that three of the most distinguished scientists of the day, Jacques-Dominique le Comte de Cassini, Pierre-François André Méchain, and Andrien-Marie Legendre, stepped ashore on Dover's stony beach.

The French contingent was met by Roy and Blagden and escorted by a troop of horse immediately to Dover Castle, the ancient fortress perched high upon the cliff tops above the town. There the three Frenchmen were able to see at first hand the flagstaff they had observed through the telescopes of their Borda circle instrument from the far side of La Manche, as the French call the English Channel.

The constable of the castle entertained the party in his luxurious apartments while his guests discussed the operation and arranged a suitable observing schedule. The French scientists were given a quantity of Roy's white lights and on 25 September, accompanied by Blagden, returned across the channel to set up their equipment. The next day, the first of the observations began. In France, the stations were sited at the church tower of Dunkerque, at Calais, and on the high cliffs at Cap Blanc Nez. In England, Dover Castle and Fairlight Down in Sussex were chosen. The trigonometrical figure these stations made is known to surveyors as a "braced quadrilateral" and is geometrically very strong. Poor visi-

Dover Castle, Kent. The flagstaff of the keep was used for the triangulation between France and England.

bility hampered the operation, and it wasn't until 17 October that the work was completed.

The power and resolution of Roy's telescope were sufficient for the English observers to make out the figures of the French party as they moved around their instruments. Once the observations were finished, the three Frenchmen returned across the channel and "repaired to London, and appear to have been highly gratified by the objects they saw, and the reception they met with in that metropolis."[16] They visited the Royal Observatory and traveled down to Slough to meet William Herschel, the discoverer of the planet Uranus. As Major Mudge recalled in 1805, "It is painful to reflect, that this is the last amicable interview which has taken place even among men of letters of the two countries; and that the hostile armies of both nations are now encamped on the very ground which was the theatre of these scientific operations."[17] He was, of course, referring to the Napoleonic Wars then raging across Europe.

It now only remained for Roy to complete the triangulation from Wrotham Hill to the coast stations and connect them with the verification baseline on Romney Marsh. This latter area was measured during the summer months by Lieutenant Fiddes of the Royal Engineers, assisted by Lieutenant Bryce of the Royal Artillery. Instead of using the delicate glass rods, Fiddes and Bryce measured a distance of 28,535.7 feet (8,698 m) with Ramsden's steel chain, from a point near Ruckinge to a spot close by the little village of Dymchurch. Unfortunately, the measurement was not without its problems, caused by a defect in the apparatus for tensioning the chain, which led to one of Roy's outbursts against his friend Jesse Ramsden; "If the maker had not been very negligent [the defect] might easily have been prevented."[18]

Apart from two outstanding stations, the triangulation was completed by the onset of winter. Over Christmas and New Year, Roy and his mathematical assistant Isaac Dalby computed the results of their labors. On 7 February 1788, Roy was pleased to write to James Lind at Windsor that they had achieved "a wonderful degree of exactness." The difference between the length of the Romney Marsh base as measured, compared with that deduced from the triangulation, was less than 1 foot. However, later, when the work was fully complete, the difference grew to 2.4 feet (73 cm), still a remarkable achievement. In late 1789, when the final results for the combined schemes of France and England were computed, Roy calculated that the difference in latitude between the observatories of Paris and Greenwich was 2°38′26″ and the longitude difference was 2°19′42″. Cassini, Legendre, and Mechain, who calculated the results independently, derived a longitude difference of 2°19′39″, just 1,200 feet (370 m) less than predicted by Nevil Maskelyne.

# 19

# GEODETIC EXPERIMENTS

"IT APPEARS THAT IN CONSEQUENCE of the late General Roy's representations in 1787, respecting the utility of the Trigonometrical Survey at that time begun in England, The East India Company very laudably; had resolved to commence a similar operation on the coast of Coromandel, or somewhere in Bengal: at the same time they intended that the length of a degree on the meridian should be determined, because a measurement of the kind had never taken place near the tropic."[1] So wrote Isaac Dalby, geodesist to the Trigonometrical Survey of Great Britain, in 1794.

The friendship that flourished between Major James Rennell and General William Roy (there were 21 years between them) began shortly after Rennell's return to London. Rennell took a house in Charles Street, Cavendish Square, a stone's throw from Roy's house in Argyll Street. Both men shared a passion for mapmaking, but whereas Rennell was inclined toward cartography and geography (for which he is celebrated as the father of the science), Roy was more inclined toward the scientific business of earth measurement, especially that of triangulation. As Rennell recovered his health (for the rest of his life he was plagued by recurring bouts of fever), he began to take a fresh interest in things scientific and was elected to the Royal Society. In 1779, he published his *Bengal Atlas* and in 1782 the celebrated *Map of Hindoostan.*

Rennell shared Roy's concern, and that of many other scientists, over the poor state of mapping in Britain. By this date, the French geodetic and topographic surveys were far in advance of anything Britain had achieved. But Britain was slowly catching up, and the early successes of the triangulation campaign led Roy and Rennell to consider that a similar operation in India would greatly benefit the mapping of the wide territories of the East India Company. In 1787 Rennell suggested that Roy write to the "honorable Company," setting out his proposals for conducting "geodetic experiments."[2]

There was never any question of going to India—Roy was far too old and far too busy. Instead Roy and Rennell turned to someone whose "qualifications for such an undertaking were undoubtedly superior to those of any other person in that quarter."[3] Their candidate was none other than the unsung hero of Schiehallion, Reuben Burrow. Roy had known Burrow from the Schiehallion days, and his opinion of the surveyor was considerably greater than that of Burrow's former employer, the refined Astronomer Royal Nevil Maskelyne. Maskelyne found relations with some of his contemporaries difficult, especially those of low birth who had not achieved "rank" through "service" or some other distinguished "attainment."

For example, William Roy, the son of a Scottish factor, had "bettered" himself through the army and had achieved the rank of general. Another of Maskelyne's friends, James Cook, was the son of a poor day laborer who through diligence and natural genius had risen to the rank of post captain in the Royal Navy. Charles Hutton, the one time pit-boy turned mathematician had succeeded in becoming a doctor of philosophy and master of the Royal Military Academy. On the other hand, Charles Mason, a baker and miller's son, was easily their equal but had not followed a military career or secured a prestigious post, so was ranked lower in the pecking order (which annoyed him). Charles Green, Maskelyne's observatory assistant before Burrow, had come in for some vitriol after he had failed to put his papers in order before inconsiderately dying of malaria on Cook's first voyage.

Maskelyne's opinion of Burrow, a tough, stocky Yorkshireman, was dismissive "on account of his inferiority of education and situation in life."[4] Burrow's crime was his truculent manner and forthrightness in dealings with his "betters." When he applied for the position of master at the Mathematical School of Christ's Hospital, the establishment stepped in to prevent it. The post went instead to William Wales, Green's brother-in-law, an extremely capable nautical mathematician and a much-loved teacher. Wales's reminiscences of his voyage with Cook to the southern ocean inspired one of his old students, the poet Samuel Taylor Coleridge, to immortalize the voyage in his most famous epic, "The Rime of the Ancient Mariner."

Despite being a bit on the rough side, Burrow was at heart a kindly man who loved his wife and family. He also loved the occasional drink and was known to be "given to pugilism." In 1780 he was back in London living with his wife Anne and their young family at 11 New Square in the Minnories, close by the Tower of London. His outwardly appearance apart, Burrow was a sensitive, well-read man, and principally self-taught. As well as Isaac Dalby, his small circle of friends at this time included the poet George Crabbe, whom he assisted with money, and fellow mathematician John Bonnycastle. This group formed a small philosophical club and met at a nearby coffee house to discourse on literature, and the sciences, and at weekends would take their families on botanical adventures. Burrow also kept a journal. It was this journal, or rather the amusing notes and acerbic observations he jotted down, that earned him an unwarranted reputation as a "scurrilous dog." On one occasion, he described William Wales as "a dirty rascal [who] carries 2 faces"[5] and his *Miscellanea Scientifica Curiosa* as nothing but "balderdash miscellany of damned stupid, ragamuffin, methodistical nonsense and spubility." Yet neither man considered the other an enemy. One encounter illustrates how this strange man attracted ill treatment and casts some light on William Wales and his mischievous friend William Bayly, Dixon's companion in Norway.

On 10 September 1775, Wales and Bayly, who had recently returned from James Cook's second voyage of discovery, dropped in to see Burrow in his new home. Both men were enjoying their celebrity status and both were well provided with cash. Their excuse for calling on Burrow was that they had a question that required his opinion; in truth, they were out for some sport. When Burrow had answered their question, Wales suggested they go to see Mr. Keech. Keech was not at home so they went to a tavern and fell to talking about the Schiehallion experiment. Bayly, who knew that Burrow despised Maskelyne, remarked innocently "that Maskelyne got great Commendation for speaking so favorably of [Burrow]" after the reading of his paper at the Royal Society, knowing full well that Burrow had not been invited to the meeting. When Burrow wondered what had possessed the astronomer royal to speak so favorably, Wales responded that he had heard Burrow "had used Maskelyne not well."[6] Burrow's reaction to this was the predictable outburst the two Williamses had contrived.

In the months following Schiehallion, Burrow satisfied himself with odd jobs—tutoring, preparing gunnery tables for "Mr. Brown at the Tower,"[7] and compiling tables of stellar aberration for Lord Charles Cavendish, Henry Cavendish's father, with whom he got on well. Among his many acquaintances at this time, Isaac Dalby was particularly close.

As luck would have it, Burrow's casual work at the Tower of London brought him into contact with Major Henry Watson (1737–1786), one-time chief engineer of Bengal who had once served under Lord Clive. In early August 1775, Burrow went to see Watson "and asked him about getting into S Sea house," East India Company's headquarters in Leadenhall Street.[8] Unfortunately, the major didn't have any suitable acquaintances at the time (he had been tarred, with some justification, with the same brush that had brought about Clive's demise) and so the matter passed. Burrow's disappointment at this, as well as the loss of the master's position to his rival Wales, was offset when Watson instead recommended him to Lord Townsend as a suitable candidate "to teach mathematics to the Cadets of the drawing room" of the tower.[9] It was a good job and grossed Burrow £100 a year, four times what he had earned at the Royal Observatory.

Reuben Burrow's time in the drawing room of the Board of Ordnance was not a complete success. His paranoia, never far away, meant he divided his colleagues into two parties—those he liked, which included Dalby, Roy, and Watson, and those he detested, which included Haines, the chief draftsman, and his deputy. Then, in 1777, his surveying talents were called upon to make "a complete Survey of the Sea Coast from the Naze in Essex to Hollesby bay in Sussex" (he meant Suffolk).[10] In this task he was assisted by cadets from the tower and from the Royal Military Academy at Woolwich. The purpose of the survey was purely defensive—to assess the long coast's vulnerability to attack by France. In 1778 or 1779, he conducted another survey, this time of "Woolwich Warren." For this work he received his normal salary but not the allowances that regular army officers got for these sorts of duties. Sometime later, he complained to the master-general of the ordnance, the duke of Richmond, that he was "never paid a farthing."[11]

By 1782, the strained relationships with Haines and with Sir Charles Frederick of the Board of Ordnance at the tower had sunk to such a low point that Burrow decided to seek alternative employment. The idea of India resurfaced, and this time Watson, now a colonel, came to the fore. Watson had recovered from his period of "disgrace" and been recalled to service in Bengal. He suggested that Burrow accompany him. It seemed a perfect solution, and Burrow accepted. He handed in his resignation on 30 April 1782 "in order to go to the East Indies." Lord Townsend suggested that Burrow write, out of courtesy, to the duke of Richmond to explain his decision to leave. It was an imprudent suggestion. Burrow certainly did write to His Grace, and in customary fashion lashed out at his enemies, recommending "superannuating" the chief draftsman and his deputy and replacing them "with Gilder and Gould," two of his friends, adding "I also recommend Mr Dalby as a proper person to fill up my place." Together with a number of other complaints and acerbic observations, the exchanges between the duke and Burrow quickly degenerated into unpleasantness of which neither man could be proud.[12] Nor did poor Dalby get the job.

With an advance of £50 from the East India Company for "necessaries" and his £100 passage money, Burrow finally left London in September and headed for Portsmouth and the *General Coote* (Captain Hoare), the East Indiaman that would take him to Bengal. As Burrow waited with the other passengers for the ship to sail, news came of the terrible tragedy of the *Royal George*. The great man o'war had capsized at Spithead, and the entire crew of 800 men and "200 whores and women" were drowned. Within days, Burrow beheld "a shocking sight" as some of the bloated bodies floated past his ship. The dead women were, apparently, ignored and allowed to rot away at sea, which made Burrow very angry indeed.

In October, after further weeks of delay because of bad weather, Reuben Burrow said his final farewells to his dear wife and his young family, and the *General Coote* set sail for Calcutta. Burrow's passage to India was as passionate in its progress as was the character of the man himself. At first, Hoare's ship had the benefit of the protection of Admiral Howe's fleet as far as the Straits of Gibraltar. Britain's Mediterranean stronghold had been under siege by the Spanish since 1779, and Howe's orders were to effect the latest attempt at relief. However, when the admiral left the *General Coote* and its escort, HMS *Bristol* (50 guns), to attend to his military duties, Burrow was incensed. He filled his journal's pages with vitriol about the admiral and his brother and their incompetence in managing the American war.

Among the passengers were 200 German mercenaries and their 9 officers. The officers (he referred to them as *Hanoverians*) occupied the great cabin, and their eating and other personal habits received special treatment in Burrow's journal. One officer in particular, by the name of Linsing, ate four times as much as any other man and became a butt of an amusing piece of mathematically lewd poetry:

> Linsing they say at Portsmouth got a whore
> And touched her up a dozen times or more.
> Should Linsing every day perform such feats
> He'd f——k but in proportion as he eats.

The German soldiers also raided the dinner table and cleared away most of the food before the other passengers could eat. This made Burrow cross; it also made two other passengers, Farringdon and Shippey, furious. The two accused Burrow of engaging in similar habits, a foolish thing to suggest to someone "given to pugalism." More seriously, Burrow crossed swords with the ship's first mate, whom he suspected, with some justification, of plotting mutiny.

This clash of personalities led the two men to settle their differences above the steerage on the quarterdeck. Our hero had the better of the sailor. The Farringdon set and the first mate then joined forces to become firm enemies of Burrow. For his pains, he was "sent to Coventry; for they would not speak with me."[13] On the other hand, the ship's captain and others of the more reasonable passengers got on well with the mathematician, and they began to eat their meals together, away from the rebels and the Germans, for the rest of the long and awful voyage.

Almost immediately upon arrival at Calcutta, Burrow wrote by way of introduction to the new governor-general of India, Maskelyne's school friend Warren Hastings. Afterward Hastings and Burrow became close acquaintances, sharing many common interests in all things Indian. As the "mathematical master to the Company's corps of engineers" at Fort William, Calcutta, the next 6 years seem to have passed without any incident of note. For the first time in his life, Burrow was well paid and holding a prestigious post, living well and sending money home, relatively content. He was no doubt enjoying the experience that so many Britons would come to know in the years of empire—a mixture of imperial superiority, a step up the rigid class ladder, and plenty of servants. He was needed and appreciated; he was free of irritating rivals; he was happy.

Apart from being an accomplished geodesist and mathematician, Reuben Burrow's interests were quite eclectic and he even taught himself Sanskrit. In Indiophiles such as Governor Hastings, Charles Wilkins, and others among the more learned Europeans, Burrow found like-minded individuals, all fascinated by the rich and varied cultures of India. In 1784, they banded together to form the Asiatic Society of Bengal and began to publish, albeit erratically, the society's journal, *Asiatic Researches*. Burrow, now a founding member of a learned society, became intrigued with Hindu mathematics and devoted much time to its study. His research into the subject and his strange, by modern standards, deduction that the Brahmins of India and the druids of Celtic Briton were somehow one and the same, was sponsored entirely by Warren Hastings.[14]

So important for history and Anglo-Indian relations, Burrow's studies became compulsory reading. As the eighth edition of the *Encyclopedia Britannica* noted, "We are indebted . . . to Mr Reuben Burrow for some of the earliest notices which reached Europe on this very curious subject. His eagerness to illustrate the history of the mathematical sciences led him to collect oriental manuscripts, some of which in the Persian language, with partial translations, were bequeathed to his friend Mr Dalby of the Royal Military College, who communicated them to those interested in the subject, about the year 1800." Unfortunately, his prowess with mathematics and historical research did not extend to home economics: "I have only 20 Gold Mohurs left out of 500 Rs [rupees] that I rec'd for the month of December because how it is gone by God I cannot tell but it is all gone in a very short time . . . I cannot make out for the blood of me. I have had no whores nor any liquor in the house."[15]

In 1788 Burrow wrote a piece for the *Asiatic Researches* on "Corrections of the Lunar Method of Finding the Longitude" and in 1790 another paper on Indian architecture entitled "Memorandums Concerning an Old Building, in the Had-jipore District, near the Gunduc River, &c." But his chief interest remained as-tronomy and geodesy.

Eighteenth-century native Indian physical astronomy was not lagging far behind that of Europe. The great astronomer-sage Maharaja Sawai Jai Singh II (1688–1743) built a number of observatories throughout India. The structures in Delhi and Jaipur, which still exist and attract thousands of tourists, were built entirely of stone and were principally used for finding the time and star posi-tions for astrological purposes. The Delhi complex contains a huge sundial with a giant "gnomon," so tall that it has stairs to climb to the top. At Jaipur, the ob-servatory has a small equatorial sundial (the Laghu Samrat Yantra), measuring time to 20 seconds, and a large hemispherical complex sundial, the Narivalaya. There is also the Jai Prakash Yantra, an armillary sphere similar to those of Eu-rope. Also at Jaipur are the so-called Zodiacal Instruments, the Rashi Yantras, used by Jai Singh to measure the changing positions of the sun and planets.

Burrow's output of papers for *Asiatic Researches* reached a peak in 1790. Four, possibly five, went to print, including "Observations of some of the Eclipses of Jupiter's Satellites" and "A Synopsis of the Different Cases that may Happen in Deducing the Longitude of One Place from Another by Means of Arnold's Chronometers, and of Finding the Rates when the Difference of Longitude is Given."[16] At some point in 1788, Burrow heard of William Roy's ideas for a geometrical survey of Bengal and discovered with considerable satisfaction that his "superior qualifications" had at last been acknowledged back home in En-gland. He, Reuben Burrow, had been specifically chosen by the greatest surveyor in Britain to conduct the "geodetic experiments." That his selection was proba-bly due more to the influence of Isaac Dalby was immaterial. Certainly his bene-factor, Colonel Watson, can have had no part in the matter because he was al-ready dead.

Roy, in making his recommendations to the East India Company for a geo-metrical survey, was expecting Burrow to implement a trigonometrical scheme along the same lines as those going forward in Britain. The resulting framework of triangles would then act as a sort of coat hanger for James Rennell's superb maps of the region. After studying Roy's work, Rennell, who as surveyor-general advocated fixing route surveys by astronomical observations, was convinced of the superiority of triangulation for mapping control.[17]

However, before any triangulation could start, Burrow had to have a baseline from which to work and a robust and accurate geographically known position from which to start. He determined to solve both requirements in one move, by measuring arcs. From his experience on Schiehallion, his time with the Board of

The Hindu observatory in Delhi.

Ordnance, and discoursing late into the night at the Swan's Nest tavern with his fellow mathematicians and astronomers ("Dalby was there"[18]), Burrow was as well acquainted with the contemporary state of knowledge as any earth measurer and knew that no meridian arc had been observed at a latitude between Bouguer's arc on the equator and Mason and Dixon's American arc at 39°12' north. Burrow resolved to fill the gap and measure one in India, on the Tropic of Cancer, 22°28' north latitude.

It will be recalled that the primary purpose of Roy's trigonometrical survey was to determine the difference in latitude and, more important, longitude between the royal observatories at Greenwich and Paris. On this matter, a letter from Dalby arrived in late 1788 with the results of the measurement. The Anglo-French scheme had shown that a difference in longitude measured through some 250 miles (400 km) of precise triangulation, when compared to that deduced from well-calibrated chronometers, was a mere 700 feet (213 m). Even more interesting for the earth measurers was that, in the link between Greenwich and Paris, they now had the length between two distant locations lying east to west rather than north to south. The importance of this piece of information was that, unlike the elliptical shape of a meridian arc, the section of the earth made by a line of latitude is a regular circle. Burrow decided that he would repeat the same method in India and include the measure of a degree of longitude with one of latitude.

Fired with enthusiasm and the prospects of greater fame, Burrow was soon to realize that his grand schemes would have to wait. The sentiments behind the East India Company's "laudable" resolve to see a trigonometrical survey in India were stronger than its finances. The pecuniary difficulties that the company was experiencing at the time, together with a good deal of doubt over the necessity of spending money on the instruments for such an obscure project, led to "some disagreement among the persons principally concerned in providing them, [and the instruments] were never sent from England."[19]

It would seem that instruments were indeed ordered, or at least enquiries about costs and delivery were made among the London makers. For advice on what to send to Burrow, the company's directors in London would have turned to William Roy and James Rennell. From his classroom in Fort William at Calcutta, Burrow first expected the instruments to arrive sometime in 1789. When nothing turned up, he sent several reminders to the directors, imploring them to secure the instruments he needed. He begged them to buy a zenith sector "at any price"; a naive request indeed to make of a hard-nosed commercial trader. He was politely informed by one of the London agents handling his affairs that an instrument was not then obtainable, but one was "expected to become available" in 1790 and would be sent to him. The inference was that an order for a specific instrument had been placed—this would probably have been in 1787—and was in the process of construction. But what was it? Perhaps a zenith sector or, maybe, a transit instrument?

If indeed the directors' had consulted Roy, Rennell, Maskelyne, or any of a dozen other scientists with interests in the proposed Indian scheme, the choice of instrument without doubt would have been a 36-inch-diameter Ramsden theodolite.

# 20

# I KNOW IT WILL ANSWER

"THE WANT OF A ZENITH SECTOR," wrote Isaac Dalby, "seems not to have discouraged him; for I find from the manuscripts which he left me . . . that in 1790 and 1791 he measured a degree of longitude, and also another of latitude under the tropic, with such instruments and other apparatus as he could procure."[1] Neither a zenith sector nor any other geodetic instrument ever arrived in Calcutta. Instead, Burrow had to make do with what he could acquire or borrow.

Dalby held his old friend's abilities with mathematics, astronomy, and surveying in the highest regard. He had no doubt that Burrow's work would prove reliable and invaluable because "a business of this kind . . . when undertaken by a person of Mr. Burrow's abilities, can hardly fail of adding to the common stock of scientific knowledge." Yet Burrow, great as his talents were, was by no means the first of his ilk to undertake geodetic experiments in the great subcontinent.

A Calcutta contemporary of Burrow, Colonel Thomas Dean Pearse (1741–1789), had been making observations for latitude and longitude along the coast and of many of the principal towns and cities since the 1770s. The East India Company provided Pearse with some fine instruments, some of which had been sent to William Hirst for the 1769 transit of Venus. These instruments included a John Ellicott astronomical regulator similar to that used by Mason and Dixon in Africa, as well as a transit and equal altitude instrument made by Jonathan Sisson. There was also a "tolerably good" Hadley quadrant of 12-inch (30-cm) radius "with two sets of divisions on the limb," made by Jesse Ramsden, and a 15-inch (38-cm) quadrant by Benjamin Martin of Charing Cross.

When in 1789 the 48-year-old Pearse died, his instruments were sold, and one or two found their way to Reuben Burrow. In addition to what he acquired from Pearse's effects, he also had a small theodolite, a sextant, a 50-foot (15.25-m) Ramsden patent steel chain supplied by the company, and several glass rods (probably recommended by Roy) from which he constructed long bamboo measuring rods. He also secured two excellent Arnold chronometers that may have come into his possession under questionable circumstances.

In the month of confusion before finally leaving England, Burrow acted as an intermediary between John Arnold (1736–1799), the London watch and chronometer maker, and Captain Hoare, of the *General Coote,* in the matter of two chronometers. Burrow collected the chronometers, each valued about £50, and brought them to the ship from London. However, it would appear that the captain may not have actually paid Arnold for the clocks before the ship's departure, or somehow the money had gone astray, and it was these "hot" items that, it would seem, Burrow had in his possession.

Burrow, being an astronomer, as well as a mathematician, had not been long in India before he realized that what was needed for safer navigation in the waters around the treacherous coasts was a modern astronomical observatory at Calcutta. In about 1790, he felt compelled to write to the company directors, proposing that they fund the building of an "Indian Greenwich." He was surprised and somewhat dismayed when the proposal was turned down flat, but not without good reason.

In 1785, an acquaintance of James Rennell, the surveyor and astronomer Captain Michael Topping (1747–1796), was sent from Britain on Alexander Dalrymple's recommendation to map the dangerous coastlines along the Coromandel. Topping has been described as "the most talented and highly qualified all round surveyor that served the East India Company during the 18th century,"[2] an accolade perhaps a shade precocious. A friend and colleague of Topping was the influential and wealthy company trader William Petrie (d. 1816). In 1786, Petrie, himself an amateur astronomer, had built a small observatory in his large back garden at Egmore, Madras, and hired the Dane Johannes Guldenheim (John Goldingham, later fellow of the Royal Society, d. 1849) as his assistant. While Petrie and Goldingham began latitude and longitude observations to determine the position of the observatory, Topping prepared to set off on his coastline surveys. Departing in November 1786, Topping and his team headed from Madras overland and spent the "cool season" in preparing maps of the rocky shores and observing latitudes and longitudes before returning to Madras in February 1787.

In January 1788, Topping headed south from Madras to survey the coasts but, before doing so, arranged for the East India Company to pay Goldingham to make observations of Jupiter's moons (for longitude) from Petrie's observatory. In 1789, when Petrie returned to Britain for a short break, Topping was put in charge of the observatory. With Petrie's support, Topping submitted a proposal to the board of directors for Petrie's observatory to become a public office, noting that "it is doubtless from considerations of this nature that the Hon'ble Court [of directors of the East India Company] have come to the resolution of thus affording their support to a science to which they are indebted for the sovereignty of a rich and extensive empire."[3]

On 19 May 1790, the directors formally accepted the idea of the observatory as "promoting the knowledge of Astronomy, Geography and Navigation in India," and thus was born the famous Madras Observatory, which did, indeed, become India's Greenwich. Two years later, Topping was appointed the company's astronomer and surveyor in command of all nonmilitary surveys; he died in 1796.

In the year Topping set off on his survey of the Indian coast, far away, in Italy, Boscovich's successors were also engaged in earth measuring. A 6.2-mile-long (10 km) baseline was measured preparatory to beginning the triangulation scheme to support a new map of Lombardy for the governor, Count Firmian of Austria. The Italians had chosen for their site the flat terrain between Somma Lombardo and Nosate. Their three iron measuring rods, engraved with the Paris *toise*, were even more elaborate than Roy's glass tubes. The rods were laid in wooden troughs so that they overlapped slightly. For fine adjustment, a special key on a toothed wheel was used to slide the rods back and forth until the graduations lined up perfectly. The baseline measurement was made in both directions, from Nosate to Somma and then back, with a difference of just two inches (5 cm).

News of the Italian work reached India just as Burrow was preparing to start work on the degree of longitude, the first made directly on the tropic of Cancer. The tropic of Cancer and its southern counterpart, the tropic of Capricorn, mark the northern and southern limits of the sun's annual passage across the equator, reaching the former at the summer solstice, about 22 June, and the latter at the winter solstice, about 22 December. The area Burrow chose for his arc measurements had to be conveniently near Calcutta so that he could continue with his other duties for the East India Company, and it had to be reasonably accessible. Fortunately, the flat expanse of the Ganges delta was ideal for baseline measurements. In the end, he chose an area lying alongside the meandering Houghli River between Chandanagar and Clives's battleground at Plassey.

It was late in the season of 1789, with the monsoon rains upon him, by the time Burrow had the route for the longitude measurement prepared and the ground cleared. Writing to his Asiatic Society colleague Sir William "Oriental" Jones, the supreme court judge of Calcutta, Burrow described his method for determining his degree of longitude: "My intention at first was to have actually measured a whole degree with rods in the same manner as others measure a base, and afterwards to determine the difference of longitude by the watches [chronometers], I have already tried it, and know it will answer; and I took some advantages by dividing the line into particular parts, &c."[4] What high court judge could these days claim to understand geography at its very cutting edge?

Burrow's only disappointment was that he was unable to secure the services of a particular assistant he required, for what reason is a mystery. As time was short before the rains came, he could only "trace out the line and secure it with bamboo pins, and measure it as exactly as possible with Ramsden's new invented chain."[5] The more delicate observations would have to wait until the next season.

For convenience's sake, Burrow divided his degree of longitude into four sections, which he roughly measured with his Ramsden chain, proposing "to measure the whole over again with rods in the cold season, as at that time two careful people and assistants may measure at the rate of two miles a day easily enough." It was hot, unpleasant, and difficult work. The rice paddies, through which the twenty-strong survey party passed, were filling up with water, which hid the treacherous bunds and flooded ditches of the paddies from the unwary. With the coming of the rains, the humidity soared toward 100 percent. Everything was sopping wet—clothing, writing paper, boots, and bedding.

By noon, the thunderclouds were boiling furiously, cutting off the sun and casting the flat, green land into somber gray shadow. The first clap of thunder sent the laborers scurrying to fabricate a makeshift shelter. They could hear the rain coming, hissing like an old steam engine. There was a sudden gust of chilly wind, and then the first fat, heavy drops began to fall. Within seconds, the landscape vanished behind a veil of torrential rain; there was more water in the atmosphere than air.

The rice paddies were perilous in other ways: the rains brought out the termites, and the toads lined up to gobble the fat insects. The toads, in turn, attracted the attention of their own predators: "I met today with an extraordinary snake," wrote Burrow in his journal, "that kills people by blowing on them, as they say. A man in leveling the ground with a cadally, cut it in two, and though there were twenty people, they all ran away from it. It was a small snake about 1½ feet long, and about a quarter of an inch thick."[6] The snake was probably *bungarus caeruleus,* the lethal Indian krait.

That rainy season, Burrow completed a treatise he had been working on for some time: "A Proof that the Hindoos Had the Bionomial Theorum." The paper attempted to demonstrate to the world that ancient Indian mathematics where far in advance of anything in the West and, through the agency of the Hindu religion, had spread far and wide. While today we might smile at some of his conclusions, at the time he penned his notions they were red-hot. He suggested,

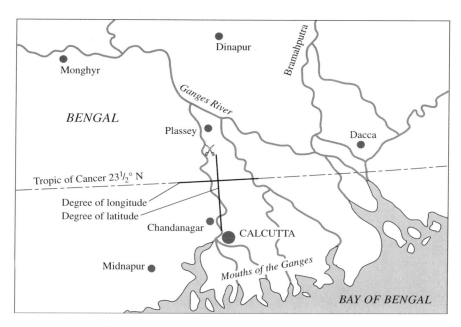

Burrow's meridian and longitude arcs in Bengal.

"from the Paradise of Moses . . . the Hindoo religion probably spread over the whole earth; there are signs of it in every northern country, and in almost every system of worship." With this paper, Burrow launched himself into the history books as one of the earliest Vendata-Perennialists and certainly one its most important founding figures, which he remains still. The rough, paranoid mathematical astronomer who had railed against Maskelyne and the stuffy English scientific establishment was gone; Burrow the Philosopher had arrived.

Returning to fieldwork, Burrow began his precise baseline measurements in April 1790 as soon as the rains ceased and the cool season began. From near "Cawksally . . . at latitude 23°29'07"N and longitude 88°19'30"E" (Navadwip), the survey crew began tracing the parallel of latitude eastward, using the theodolite to keep on the line. A line of latitude when set out on the ground forms a very shallow curve and not, as might at first be thought, a straight line. To establish the correct direction for the baseline, every night Burrow observed an azimuth (angle from true north) from the Pole Star "at elongation," just as Mason and Dixon had when setting out the border separating Maryland from Pennsylvania.

The measurements across the flooded paddies were made with the Ramsden 50-foot chain. "Here the rice fields were full of water, and the chain was laid constantly within a finger's breadth of the water,"[7] wrote Burrow, recording the project's progress. Exactly how he did this is a matter of speculation. Probably he was being smart and using the level surface of the floodwater to keep his measuring apparatus horizontal. The tenacious Yorkshireman would have accomplished this by using planks or troughs supported on stands to carry the chain, just as he had at Schiehallion. The process was exhausting; the continual soaking and the perils of water full of leaches and hidden ditches made the business of accurate measurement extremely difficult and frustrating. But even the thick groves of bush and snake-infested bamboo breaks didn't sap Burrow's newfound joy in life and keep him from making jokes: "Why is a Bengal marsh like Ireland? Because it produces Paddy."[8]

By the middle of May, Burrow and his team of Bengalis had measured some 33 miles east from Cawksally, a good average of nearly a mile a day. In June, when much of the water had drained away to the rivers, he returned to the starting point and began working westward. By the time the rains returned in the middle of July, he had measured another 15.5 miles.

The arrival of the monsoon, which brought an end to fieldwork, coincided with the arrival of his wife, Anne, and their son and three daughters. It must have been a very happy reunion after such a long separation. Anne had only 5 months with her husband before the onset of the cool season, and he had to return to the business of measurement and astronomical observation. From December 1790, they pushed the baseline west another 24.5 miles, reaching Dhoraparah (modern Guskhara) on 22 January 1791.

This last phase of the measurement was accomplished by using a combination of chaining and the bamboo rods. Throughout the long, tedious process, the chain and the rods were always carefully calibrated against Ramsden's brass scale standard, and any differences were carefully noted. The final value for the east-west baseline was calculated to be 212,670 feet, or 40.28 miles (64.82 km). All that remained was to determine the actual difference in longitude between the terminal points. For this, Burrow used nine chronometers, the two Arnold units of his own supplemented by another seven that he had borrowed from sailing masters of the company's ships. Between April and May 1791, he made twelve cross-country journeys between Cawksally and Dhoraparah, measuring the apparent local time at each place and comparing these figures against the chronometers.

The usual method of doing so was with the astronomical quadrant, observing "equal altitudes" of the sun or stars. Stellar bodies reached their maximum altitude (vertical angle) above the horizon the moment they crossed the north-south line of the meridian. Because it is almost impossible to detect or measure this moment with any pretence to accuracy, the technique employed was to observe equal altitudes. As the name suggests, the altitude of a star was measured shortly before it crossed the meridian, and the time from the chronometer was noted. The angle on the quadrant was then kept fixed, and at the precise moment the star again made the same angle to the horizon—the equal altitude— the time was again noted. The mean of the two times was the precise instant the star crossed the meridian.

By this method, the mean of all nine watches was "2m 32s for the difference of longitude, answering to the apparent length 212,670 feet."[9] The final result for Burrow's degree of longitude (calculated some years later by Isaac Dalby) was 335,937 feet, or 63.62 miles (Dalby made a small error, and the distance should have been 335,795 ft, or 102.35 km).

In the new year of 1791, Burrow began the measurement of his arc of the meridian "in latitude 22°44′12″,7 at Poal, or Pole, a place nearly under the meridian of Cawksally."[10] The first task was to establish the difference in latitude between Poal and Abadanga, the proposed northern terminal. In a letter to Edmund Morris, another member of the Asiatic Society, dated 30 January 1791, Burrow wrote, "I have finished the measurement of the degree of longitude, and shall not be long about the degree of latitude. The astronomical part is already pretty forward. I hope there will be a third volume of the Asiatic Researches, as the two measurements will make an useful article in it." Using the 12-inch astronomical quadrant, Burrow observed 59 latitudes at Poal and another 115 at Abadanga. From a mean of the latitudes, the difference from Poal to Abadanga was calculated to be 1°07′59″. With the astronomical observations out of the way, Burrow was ready to start on the physical distance measurement.

This measurement was taken directly north in the meridian "to Abadanga in latitude 23°52′11″,7."[11] The character of the land across which he measured was much as Mason and Dixon had experienced in Delaware: "the ground was level as the sea." From a comment in his report, it would appear at first that the rods he used for the measure must have been very long: "Feb. 15, I measured the bamboo, and found it 194 feet 4 inches, it therefore had stretched."[12] Whether he was being economical and had lined up the bamboo rods end to end, measuring the whole against the chain, or had in fact constructed a very long rod is a puzzle. A more plausible explanation is that the note is simply a mistake. Rods were typically cut to approximately 20 feet, and what in fact Burrow meant to write was 19 feet 4 inches (there is among the notes a clear example of this simple sort of mistake) or, even more likely, it is a printer's error.

The first 19 miles of the meridian line was measured twice over, and the two distances differed by 9 feet, or just over a 2-foot error in each mile measured.

After adjusting his bamboo rods for the effects of temperature and humidity, Burrow found the distance from Poal to Abadanga was 411,004 feet, or 77.84 miles (125.27 km). The work done, it was time to retire for the wet season.

From the meridian arc distance and the measured difference in latitude, Isaac Dalby was later able to calculate that "as 1°07'59":411,004 feet:: 1° : 60,457 fath. for 1° on the meridian in latitude 23°18', the middle latitude nearly, between Poal and Abadanga."[13] The length of Burrow's meridian degree on the tropic of Cancer was 362,742 feet, or 68.70 miles (105.64 km).

At the time of Burrow's observations in India, there was wide agreement on the earth's dimensions and its shape, but there still remained the enigma of La Caillé's meridian arc at the Cape of Good Hope. This measurement appeared to suggest that the earth had a prolate form south of the equator. However, from Burrow's longitude and latitude arcs in Bengal, in the words of Isaac Dalby, "it seems extremely probable, that the meridians are not elliptical in low latitudes; but that the earth (as M. Bouguer supposed) is flatter in a north and south direction, or more of a globular form in those parts, than an ellipsoid. And were we certain how far M. de la Caille's degree at the Cape of Good Hope is erroneous (it being longer than the measured degree in lat. 45°N) it might serve to shew if the variation from an ellipsoid is greater, or extends farther on the south side of the equator, than on the north."[14]

For the purpose of determining degrees of arc for the shape of the earth the quality of the two Indian baselines was barely acceptable. The result, which Dalby calculated, appears to show that Burrow's attempts at calibrating his measuring rods underestimated the havoc that the climate wreaked on the bamboo. Yet, given the fact that the tools he had to measure the latitude were woefully inferior to those of his contemporaries, the meridian arc was only 600 feet (180 m) short of the distance, as Everest determined it many years later.

For a formal triangulation scheme, the lines were hopelessly inadequate as baselines, and this Burrow acknowledged. At 40 miles and 78 miles, respectively, they were also far too long. We can only speculate about his next move; in all probability, he planned to remeasure the section of the Poal-to-Abadanga line that lay nearest to Calcutta and reobserve the latitudes. In Ramsden's patent steel chain he had the measuring tool, although not the calibration standard, and its condition by this time must have been a concern. But of the most vital piece of equipment he needed, the promised "instrument for 1790," there was no sign.

As the monsoon and miserable weather cleared away, Rueben Burrow left Calcutta to go to the northern province of Bihar. The purpose of the journey may have been something of a reconnaissance or, more likely, some rest and recuperation with his family. Let us hope that his wife and family went along, for the final tragedy was at hand. A faltering constitution brought on by the exertions of the survey and the cruel climate made him vulnerable to malaria fever. On 7 June 1792, at the rest house at Buxor, Rueben Burrow died. He was just 45 years old.

Although robbed by nature of his chance to undertake the first triangulation scheme in India, Reuben Burrow's arcs on the tropic of Cancer added significantly to the growing volumes of knowledge on the shape of the earth. His labors were the genesis for the great surveys of India that were to follow. Thwarted by Nevil Maskelyne of any honor for the Schiehallion experiment, Burrow cannot be denied the credit for measuring the first Indian arc.

# 21

## OFFERING VIOLENCE TO NATURE

WILLIAM ROY RETURNED TO LONDON from his exertions on the British trigonometrical survey in declining health. Much of his time was taken up with Dalby in completing the hundreds of computations and preparing reports. In November 1789, the ailing surveyor, now approaching 64, went to Lisbon in hopes of restoring his strength.

He was back in Argyll Street the following April and began immediately to prepare his report for presentation to the Royal Society. Despite a winter break in the sun, his health continued to deteriorate and adversely affect his temper. The draft of his report on the triangulation, read to a committee of the Royal Society in his absence, contained a number of very derogatory remarks about Jesse Ramsden and his tardiness and, worse, challenging the quality of his extraordinary workmanship.

These slights were too much for the aging Ramsden, and many of the fellows of the society concurred. Ramsden felt obliged to write a strong rebuke; he was upset and deeply hurt by Roy's remarks, as he had, despite their disagreements and all the delays, always regarded Roy as someone "with whom I considered myself in Friendship."[1] Without any doubt, had Roy been in better health and mentally robust, he would never have dreamed of humiliating the old instrument maker in such a public way.

Ramsden petitioned the council to strike out the offending sections before the paper was printed in the society's *Philosophical Transactions*.[2] This, William Roy's last paper, was published posthumously, with the offending passages expunged, as Roy would certainly have wished. In the early hours of 1 July 1790, at the age of 64, the great surveyor died.

Roy's death was a great loss to his science and left behind a vacuum difficult to fill. For a while, hopes for a continuance of Britain's national survey seemed to fade. The duke of Richmond, master-general of the ordnance, was only too aware of Roy's aspirations for a national survey, and he was not about to let his old friend's noble and militarily important ideas wither on the vine. If Britain were to go to war with France, good maps of the nation's southern coasts and counties would be imperative for the defense of the realm.

Since the beginning of the eighteenth-century, the nature of warfare had changed dramatically and its technologies had greatly advanced. Artillery had developed steadily from a form of destructive art toward a more scientific form of devastation. Ship design, too, was much improved, and naval vessels were larger, faster, and more powerful. Military tactics had changed, and the old, trusted political alliance system, encapsulated in the principle of the balance of power, was fast failing. Then, in 1789, France's infamous Bastille prison fell to an assault by the Paris mob, marking the start of the French Revolution. Rumors of war were in the air, and fear ran through the establishment classes of a similar uprising in Britain.

The social unrest and political bloodletting taking place just 20 miles away across the Straits of Dover had its admirers in England, who saw the events as a social revolt, the politicizing of Engel's "veritable storm." Others were supremely confident in the strength of fortress Britain. Writing in the early days of the Napoleonic conflict, William Mudge happily commented, "The least breadth of the channel, therefore, does not exceed twenty miles—a narrow but strong barrier—one of those indelible lines which nature has kindly traced out on the surface of the earth to resist the ambition and preserve the independence of nations."[3] But not everyone agreed with the major's assessment. The old barriers, physical and metaphorical, were beginning to fail, and the ruling classes of Britain were, with some justification, concerned that an invasion by France was inevitable.

New, modern, and robust fortifications—vastly improved designs of the versions first made famous by Marshal Vauban in the early years of the century—were constructed. New theories of fortification, postulated in particular by the ingenious French engineer Marc-René de Montalembert, were further advanced in the building of ramparts bristling with batteries of cannon set in huge impregnable *casemates*. British military engineers lived in a French world of *bastions,* battery *cavaliers, tenailles* and *caponiers, glacis* and *lunettes;* this was the eighteenth-century equivalent of Star Wars. In 1756, the military town of Chatham, in Kent, guarding the royal dockyard that had been violated by the Dutch 100 years earlier, was strengthened with a bastioned stone masonry line and new, deep-set fortifications. The important port of Sheerness, guarding the Thames estuary, was likewise strengthened in the 1780s. The already frowning eminence of Dover Castle, "the gateway to England," was radically altered from a medieval fortress into a modern military stronghold, and across the valley from the castle new heavy fortifications were being thrown up.[4] Accurate maps were increasingly an essential part of the warmongers' arsenal.

With these military preparations in mind and an eye to the defense of the realm, the duke of Richmond consulted his colleagues in the Royal Society on the best way forward for the national survey. Until then, the society had accepted responsibility for the work but was keen to offload its costly liability. After all, it was a learned institution and not a military department. In the event, it was agreed that the duke's Board of Ordnance should take over the work of the survey entirely and that the Royal Society would continue to lend its assistance and advice. Ever a quasi-army operation, the British survey was about to be formally militarized.

The duke approached Charles Hutton (mathematician of Schiehallion fame), the master of the Royal Military College at Woolwich, for suggestions on who might take charge of the national survey. Isaac Dalby, the only survivor still active from the days of William Roy, was an aging civilian, and that would never do. Hutton had a lot of respect for Dalby both as a man and mathematician, and it can be safely assumed that he consulted him first. In the event he recommended to the duke that Major Edward Williams and Lieutenant William Mudge of the Royal Artillery were "the best mathematicians in the two Corps, and the fittest officers for this duty."[5]

The duke accepted Hutton's endorsements, and the two officers were duly appointed. On 10 July 1791, they were issued with the instructions to "carry on the Trigonometrical Survey with the assistance of Mr Dalby." This date is generally considered to be the founding of Britain's national mapping organization, the Ordnance Survey, although its title was never adopted as such until the 1830s, when the name was accepted by default.

At about this time "a casual opportunity presented itself to the Duke of Richmond of purchasing a very fine instrument, the workmanship of Mr Ramsden, of a similar construction to that which was used by General Roy, but with some improvements."[6] This instrument was another 36-inch diameter theodolite and, from all reliable accounts, had been ordered by the East India Company for its surveys in India; for some reason, the company had decided it was no longer needed.

The official records surrounding this instrument are vague and contradictory; what is certain is that the duke, when he heard of its availability, immediately arranged to acquire it. The instrument, now on display in London's Science Museum, is stamped with the date 1791, the year of purchase, not of construction. Without doubt, it can be none other than the instrument promised to Reuben Burrow for 1790. That an instrument existed is testified by Mudge's account of the events, and after all, if anyone knew it would have been Mudge: "In the end,

Ramsden's second 36-inch theodolite of 1791 built for the East
India Company and sold to the Board of Ordnance.
Courtesy Ordnance Survey. © Crown Copyright.

some misunderstanding arose; and in a fit of ill-humour, or of ill-timed economy, induced the sovereigns of India to refuse an instrument which could do
nothing to enlarge their dominions, though in skilful hands it might have done
much to render them more secure."[7] Had Burrow not died and had the theodolite been delivered to his "skilful hands," the story of the survey of India, and
that of mapping science itself, might have been very different, as well as advanced by at least 20 years.

The headquarters for the Board of Ordnance was at the Tower of London,
and therefore it was logical that the duke's new mapping department was based
at the same location. The survey offices were located in an old armory adjacent

to the Bowyer Tower, to the north of the White Tower, William I's great stone "donjon." Director Williams, Lieutenant Mudge, and Isaac Dalby were soon joined by James Gardner, a man with a fine sense of humor, as the chief drafts-man. A working party of artillerymen was also provided to act as laborers and surveying assistants.

Apart from their shiny new theodolite, the surveyors also had two new steel chains and some lesser instruments, all constructed at Jesse Ramsden's bustling workshops at 199 Piccadilly. For some unknown reason, the first Ramsden theodo-lite used by Roy and belonging to the Royal Society was not included in the in-ventory. There is more than a suspicion that Mudge and Williams did not al-ways see eye to eye and that, when Mudge recommended acquiring the society's instrument, Williams vetoed the idea.

In fact, there is considerable doubt that Williams was ever anything more than a figurehead; he does not feature at all in Mudge and Dalby's 1799–1805 ac-count of the trigonometrical survey: "In 1791, Captain Mudge of the Royal Artillery, and Mr Dalby . . . both well qualified for the work they were to under-take, had the care of the trigonometrical survey committed to them, and re-ceived their instructions from the Master-General of the Ordnance."[8] The char-itable suggestion is that Williams was in poor health and this prevented him from participation in the survey. In any event, the man died in 1798; therefore, the first director is, at best, a shadowy figure.[9]

The first task ordained for the newly formed Board of Ordnance's survey was to remeasure the base on Hounslow Heath with Ramsden's new, improved chains. On 15 August 1791, at the request of the duke of Richmond, Sir Joseph Banks, Nevil Maskelyne, Charles Hutton, and Jesse Ramsden joined the three survey officers and their artillerymen on Hounslow Heath. Ramsden's chains were laid out within the same wooden coffers that had been used to support Roy's glass rods. These, in turn, were set on trestles to keep them clear of the ground and perfectly horizontal. The measurement was made in both directions, and the average was found to be just 2¾-inches (7-cm) longer than Roy's mea-sure by the glass rods. The final value for the baseline was averaged between the 1784 and 1791 values and agreed by all concerned to be 27,404.2 feet (8,352.8 m).

Over the winter period, Mudge and Dalby consulted with the duke of Rich-mond about their plans for the survey. The priority, as ever, was defense, and it was agreed that the triangulation should be extended expeditiously to the coasts of Kent and Sussex. With the coming of spring 1792, Mudge and Dalby set off on a reconnaissance of the land to identify and select the places where they would set up the great theodolite. Instead of using Roy's white light pyrotech-nics, specially improved observing lamps were ordered from Howard's of Old

Street, London. Observations began from the Hounslow base in the summer and continued until the bad weather arrived.

Early the next year, another "base of verification" was selected on Salisbury Plain in Wiltshire between Beacon Hill, near Amesbury, and the ruined castle of Old Sarum, not far from Salisbury. Apart from the interruption necessary to measure this base in 1794, the triangulation proceeded westward until, by the close of 1795, a chain of triangles extended from Dover all the way to Land's End at the southwestern extremity of England. The Salisbury Plain baseline was measured by the same method as the Hounslow Heath base, using Ramsden's chains. Its length by the calibrated chains was found to be 36,574.4 feet (11,147.9 m). The length, calculated from the triangles linking it to Hounslow, was just 1 inch shorter.

In 1797, Mudge and Dalby extended their triangulation across the ocean from Land's End to the Scilly Isles. When Dalby computed the results, he thought at first there was some mistake in his mathematics. The numbers were recomputed and checked by Mudge. There was no error: the Scilly Isles were not where they were supposed to be. The error in position was 26'37", and the islands actually lay some 20 nautical miles (37 km) from where the Admiralty's charts placed them.[10]

It is, then, ironic that the loss of Sir Cloudesley Shovell (1650–1707) and his naval squadron, returning from victory over the Spanish and French fleet at Toulon (the event that led to Queen Anne's announcement of the £20,000 prize for the first person to solve the longitude problem), should have taken place on these very islands and that the error had remained uncorrected for nearly 100 years.[11]

The duke of Richmond retired from active service in February 1795, and his place as master-general of the ordnance passed to Lord Cornwallis, the general who had surrendered Yorktown and lost for Britain its American colonies. In the same year, the work of the survey was augmented with the establishment of a Corps of Surveyors. These men were under the direct jurisdiction of the Board of Ordnance and were supervised by Chief Draftsman James Gardner. Their function was to do the actual "topographical" mapping of the land, recording all its features and significant detail at a scale of 2 inches to the mile.

A new 18-inch (46-cm) Ramsden theodolite was procured and used by Gardner to observe and fix "in fill" points between the main triangulation. The Corps of Surveyors then used these "tertiary" control points as the basis for their mapping by compass and perambulator. Why the corps never resorted to the plane table, the ancient but perfect surveyors' tool for small-scale mapping, is a mystery. Even in 1926, when the rest of Britain's Empire and dependencies

William Mudge.
Courtesy Ordnance Survey.
© Crown Copyright.

were happily plane tabling away, the Ordnance Survey[12] steadfastly shunned the expedient device. It was only after World War II that the survey finally accepted that the plane table was, after all, a useful tool for surveying.

In 1798, the absentee landlord of the survey, Director Edward Williams, died and the vacancy passed uncontested to Major Mudge. William Mudge (1762–1820) was born in the naval town of Plymouth, the son of Dr. John Mudge and godson of his father's friend Samuel Johnson. At the age of 15, he was enrolled at the Royal Military College in Woolwich. In 1779 he got his promotion to second lieutenant in the Royal Artillery and was sent to South Carolina to fight in the American Revolution under Cornwallis. Returning to England at the end of the hostilities, young Mudge was posted to the Tower of London, where he studied higher mathematics under Charles Hutton and Reuben Burrow. On his appointment as director of the survey, he was elected a fellow of the Royal Society.

Of an "admirable disposition and even temper," the new director quickly secured the Ramsden theodolite used by Roy. For 11 years, the great instrument had been gathering dust in the basement of the Royal Society. Ramsden was commissioned to refurbish the instrument and to fit new microscopes for reading the circles. Mudge realized that acquiring the instrument and augmenting the survey with additional resources would be beneficial to the business "were the survey prosecuted on a more extensive scale."[13]

The Royal Society of Art's £100 prize had stimulated much mapmaking of England's counties, and it was envisaged that these maps would be incorporated into the general scheme of a unified map of the country. Isaac Taylor had surveyed Dorset, and Benjamin Donn had won a prize for his map of Devon. Likewise, maps of Derbyshire, Northumberland, Leicestershire, Somerset, Suffolk, Lancashire, Hampshire, and Sussex were all published by 1796.[14] However, rectifying, or at least matching, these county maps to the Board of Ordnance's new triangulation stations showed up many serious faults. Errors of as much as 10 percent in the distances between major towns were not uncommon. The inadequacy of the mapping had stimulated the president of the Royal Society, Sir Joseph Banks, to compare Britain's maps with those produced by Major Rennell of India, whereupon he expressed the forlorn wish of all Britons that they "could boast a general map of their island as well executed."

In his *Account of the Survey,* Mudge wrote of "the very erroneous state of our maps," citing in particular the maps of Kent, Dorset, and Devon (which had won a prize). These were all key military areas. Mudge's conclusions for correcting this unacceptable state of affairs, of leaving mapping entirely to the private sector, was to advocate "some general map, published on the same principle with the Carte de France, a performance highly celebrated."[15] In fact, Mudge had already prepared the ground and had struck a deal with London's leading cartographer, William Faden (1749–1836), geographer to King George, to publish a map of Kent based on the survey's triangulation network and the topographical survey work of the Corps of Surveyors.

As with many things in the eighteenth century, the borderline between private and public initiatives is blurred. Faden, who had taken over the business from Thomas Jeffreys in 1771, actively supported the embryonic survey in many ways, including preparing and publishing the *Accounts of the Survey,* no doubt realizing the pecuniary benefits such a liaison would bring him. Faden's 1-inch map of Kent, dedicated "to the Master-General and Honourable Board of Ordnance," was published in 1801 and was the very first ordnance map.

In 1795, not to be outdone by the army and the Board of Ordnance's survey office (and years after France had established its own department), the British Admiralty decided to establish a Hydrographic Office to collect and produce long-overdue sea charts for the Royal Navy. Whereas the masters of the East India Company's ships were equipped with fine sea charts, the navy's sailing masters and navigators had to acquire their own. Years earlier, King George III had decided to make James Cook the Admiralty's hydrographer on the explorer's return from his third voyage of discovery. Cook's unfortunate death and consumption by cannibalistic Hawaiians in 1779 delayed any decision on the matter for 16 years. When the lords of the admiralty finally made up their minds, their choice of director for the new department was 58-year-old Alexander Dalrymple, the man they had firmly rejected for command of the expedition to the South Seas in 1768.

James Rennell's friend and a friend of the influential Joseph Banks, Dalrymple had held a similar post since 1779 as the first hydrographer of the East India Company. Even before he returned to London from India in 1765, Dalrymple had been collecting maritime information—charts and sketches from sailors around the world—and he was an acknowledged authority on marine matters. He believed firmly in the existence of a giant land mass in the far south, reasoning according to the old Aristotelian argument that "an imbalance of the known oceanic mass with the known land mass in the southern regions by a ration of about 8:1, indicated a large continent to be found."[16]

As a hydrographer, Dalrymple was prodigious in producing over 1,000 sea charts. His harmless fixation with the "great southern continent" is reflected in his Pacific charts long after his adversary James Cook had shown that no such continent existed. By the close of the century, Dalrymple's charts, Rennell's work on ocean currents, Maskelyne's "lunar distance," Harrison's expensive chronometer for finding the longitude, and John Bird's affordable sextant, were saving thousands of lives and millions of tons of cargo.

In 1800, old Jesse Ramsden died, just a month before his last magnificent instrumental creation was delivered to the Board of Ordnance. This was a remarkable 8-foot zenith sector, begun in 1795, which William Mudge intended to use for measuring the first meridian arc undertaken in Britain. All the science and learning of the earth measurers during the preceding 100 years went into the design of the British meridian arc. Mudge chose the line of his meridian with the greatest care so that it would be "free, as far as could be foreseen, from the action of any disturbing forces."[17] Anxious to avoid the worst excesses of the attraction of mountains, he selected Dunnose, on the Isle of Wight, as his southern starting point and Clifton, near Doncaster, as the northern terminal. The 196-mile (315-km) line joining these two places was, as far as Mudge could judge, free of any significant hills and valleys that could generate enough gravity to deflect the plumb line.

To determine the ground distance between the two points, Mudge observed a new series of triangles linking Dunnose and Clifton with the baselines on

Hounslow Heath and Salisbury Plain. To ensure that the chain of triangles closed satisfactorily, an additional 5-mile-long "base of verification" was measured across remote Misterton Carr, not far from Clifton. This work occupied him from 1800 to 1801.

Some years before, Mudge had determined the length of a degree of longitude by using the new triangulation scheme spanning southern England to calculate the distance between Dunnose eastward to Beachy Head in Sussex, the highest cliff face along the coast of southern Britain. When the longitude between the two points was computed and compared with the predicted distance, calculated according to what it should have been, Mudge had a nasty shock. He had steadfastly believed in the oblate spheroid as being the true shape for the earth, but his calculations appeared to suggest that southern England was far more oblate than it should have been; it bulged out too much. In fact, it was the same problem as La Caillé had experienced in South Africa. His conclusion was that the attraction of mountains or, as there were no mountains in southern England, local causes "may perhaps affect the direction of gravity."[18]

This conclusion led him and others to conjecture (more or less rightly, as it turned out), that "the figure of the Earth is not a solid formed by the revolution of an ellipse on its axis; and the agreement or disagreement of the measures of degrees with one another, is not to be judged of by their agreement or disagreement with this hypothesis. To attempt to judge of them in that manner, may be offering violence to nature, and may be only trying to reconcile her phenomena with out conjectural or arbitrary theories."[19] Mudge hoped that his discoveries were aberrations and that the proposed 196-mile-long meridian arc from Dunnose to Clifton would cast better light on the matter.

As soon as the new zenith sector arrived from its maker's workshop, it was taken to Greenwich to be calibrated against the observatory's instrument. The aging Nevil Maskelyne took it upon himself to instruct Mudge on how to use the instrument to best effect and how to reduce the latitude observations and apply the necessary astronomical corrections. On 9 May 1802, the 8-foot-long telescope of the sector was set up in its "strong pyramidal frame" amid the ancient burial grounds of chalky Bonchurch Down near Dunnose Point, and observations were begun.

The weather delays that had plagued Maskelyne on Schiehallion were reduced to minor inconveniences in the more temperate southern climate. By June the work on the Isle of Wight was complete, and the survey party packed up the instruments and headed north. By the end of August 1802, the observations at Beacon Hill, on the outskirts of the little village of Clifton, were also complete. All that remained was the mid-point latitude observations to be taken at Arbury Hill in Northamptonshire. Mudge was back in his office at the Tower of London by the end of October and ready to calculate the results.

The difference in latitude between Dunnose and Clifton was found to be 2°50′23.38″, and the distance separating the two points was calculated from the network of triangles to be 196.29 miles (315.9 km). This gave the length of a degree of latitude at the midpoint of 69.114 miles (111.2 km), which seemed to Mudge to be a touch on the short side. The really disturbing results though came when he computed the length of a degree of latitude for the midpoint between Dunnose and Arbury and between Arbury and Clifton. Theoretically, the ground length of the degrees should have been getting longer toward the north, but in fact Mudge found that they were instead getting shorter.

He introduced other observations to his calculations, those taken by the duke of Marlborough at Blenheim Palace and those recorded by Maskelyne at Greenwich Observatory. The results were the same—the degrees were definitely getting shorter rather than longer. "These inconstancies are very striking," he wrote, "when it is considered that, on the supposition of the earth being compressed at the poles, the degrees of the meridian must go on increasing at a uniform rate, each degree exceeding that immediately to the south of it by about twenty fathoms."[20]

Mudge's faith in an oblate spheroid was unshakable, and he was certain that it was gravity's tricks that were corrupting his observations, even in the absence of any mountains. Mudge was mystified, but he never doubted that Maskelyne and his fellow academicians around the world were correct and that his peculiar results were the consequences of gravity anomalies. The plumb line, he was certain, had been "drawn towards the south at all the stations, and that by attractive forces." Unfortunately, for the time being, there was nothing more he could do.

# 22

## A MERITORIOUS FOREIGNER

IT HAD BEEN A MATTER OF just 70 hectic years between the first major expedition to measure the shape of the earth and Mudge's arc measure. In that short time, France, Britain, and Russia—and, to a lesser extent, Germany, Holland, Italy, and Scandinavia—had developed the finest measuring instruments the world had even seen and had evolved methods and ideas that in the previous century would have been unimaginable, even blasphemous.

France and Britain used the new discoveries and the new technology to map their burgeoning overseas possessions, and across the ocean, in the New World, the newborn United States of America was coming to terms with the consequences of its independence. The land over which its federalist government had dominion was already vast, half as large again as it had been at the end of the War of Independence. And now, at the end of the old century, the United States was poised to more than double in size and occupy a slice of the continent that would span 2,300 miles and extend 1,300 miles from temperate north to the subtropical south. The ubiquitous gridiron monotony of its land system was already in the making, and at a local level, surveying and mapmaking were of a high proficiency. However, at a national level, there was no system for providing a consistent map for the young country or a means to control the exploitation of its extensive lands. This was about to change.

Jesse Ramsden's first great theodolite of 1787, built for the Royal Society and used by William Roy and his successors, provided nearly 60 years of continuous service until, in 1853, it was finally put on the retired list. During World War II, the giant theodolite was utterly destroyed when the Ordnance Survey's offices in Southampton were bombed and burned during an air raid, although its twin rests safely in London's Science Museum.

There was another one, however. In 1797, Jesse Ramsden delivered, 3 years late, the third of his 3-foot theodolites to the Swiss Canton of Berne for use on the survey of Switzerland. The survey was being conducted by Johann Tralles and a 27-year-old with the romantic name of Ferdinand Rudolf Hassler von Aarau (1770–1843). Whereas his countryman des Barres (who was still alive), became a naturalized Briton, Hassler was to become a naturalized citizen of the United States of America and would go on to great things indeed.

Ferdinand Hassler was born in Aarau, Switzerland, the son of wealthy artisans in the watch business. In 1786, Hassler's father secured for his son, at the age of only 16, a position in the Berne revenue bureau. It was in the revenue service that he first learned surveying, particularly of property and public lands for tax levies. His work as a surveyor was interspersed with study at the University of Berne, where he studied law and politics preparatory to a full-time and comfortable career as a civil servant. It was at the university that he met Johann Georg Tralles (1763–1822), a German mathematician, physicist, and savant from Hamburg, whose skill, expertise, and personality so profoundly affected the young Hassler that he abandoned his law studies and instead focused his considerable mental energy on the arts of mapping. Tralles, for his part, saw great potential in the eager youth, and he personally instructed his young protégé in the mysteries of higher mathematics, astronomy, and earth measuring.

In 1791, Hassler and Tralles measured a 7.6-mile-long baseline at Aarberg, near Berne, using a set of Jesse Ramsden's precision chains.[1] That same year, Hassler's wealthy father sponsored his son and his tutor to perform an experiment with a small triangulation network. The absence of maps of Switzerland and the results of his trigonometrical experiments decided Tralles to propose to the Economic Society of Berne a plan for a geodetic survey of the canton.

A board of commission was set up, and unsurprisingly, its members concluded that no adequate map existed "of Switzerland in general and the Canton of Berne in particular, notwithstanding the numerous drawings which bear the name of land maps."[2] The commissioners were satisfied with Tralles's proposal. They recorded that "Professor Tralles has shown the last few years by measurements and observations of a degree of precision not reached by others that not even the latitude of the capital city of Berne or of any other place in the Canton had up to this time been determined accurately."

With a favorable report and sufficient funding, the canton of Berne immediately commissioned the two men to put their proposals into effect. Tralles, conversant with Maskelyne's gravity experiment on Schiehallion, a peak of a mere few thousand feet, reasoned that the somewhat loftier Swiss Alps would exert a significant effect because of the attraction of mountains. If ignored, the gravity anomaly would prejudice the accuracy of the survey, detracting from the commission's expectations of securing "the thanks of the world of learning and be a lasting honour" to the glory of Swiss science.[3]

Hassler and Tralles's trigonometrical survey of the canton occupied a leisurely 5 years, during which time they endeavored to collect as much information about their science as they could. Through correspondence and direct contact, Hassler and Tralles were familiar figures among the philosophers of Europe. In 1793, Hassler visited revolutionary Paris, where he met Jean-Baptiste Delambre, then preparing for the measurement of a new Franco-Spanish meridian arc; he examined Charles de Borda's circle and his *règles,* the precision measuring rods the chevalier had designed. He also met with the Parisian instrument maker Etienne Lenoir and with the leading mathematical savant, Jérôme Lalande, who loaned Hassler a copy of the French official toise for the Swiss survey.[4]

But it was for the trigonometrical measurements that the canton's commissioners acquired the third Ramsden theodolite. None other than Sir Joseph Banks himself, in London, agreed to handle negotiations for the Swiss while supervising the instrument's construction. As Roy had discovered to his consternation, Ramsden was not renowned for keeping to his promised times; however, in this instance, the instrument maker had the benefit of war, with its uncertain political climate, to blame for the delay. To get the 200-pound (91-kg) theodolite through the numerous checkpoints across turbulent France and into Switzerland, Tralles was obliged to break the instrument into its constituent parts to send them through the lines, passed off as harmless scientific spare parts. The ruse worked; the instrument's sundry parts arrived safely in Berne.

The triangulation scheme planned by Tralles and Hassler consisted of twenty-five stations. As ever, the work began with the measuring of a precise baseline, in this instance a remeasure of the old Aarberg base. The latitudes at each end of the base were observed to gauge what effect the mountains might be having. Hardly had the work begun when all their plans were cast into confusion. From across the border, the French Revolution marched into Switzerland to spread its *fraternité* among the little country's people, which established the new Republique Helvetien (1798–1804).

As in France, so in Switzerland: the administration and fabric of society were thrown into chaos. The old regime was unceremoniously discarded, its members dismissed, and a new revolutionary government established. Old cantons were restructured; new ones were formed, their administrations subservient to a centralized French model government. Hassler was careful not to be drawn into the melée, wise enough not to support the Swiss revolutionaries or their politicking, yet shrewd enough to go along with events. He was a reluctant revolutionary, being "so situated, that being called in consultation, when no government actually existed, it happened that I penned the declaration of separation of the new canton formed around my native place from that of Berne, with which it had been united."[5]

In 1798, Hassler married Marianne Gaillard, the socialite daughter of a Bernese teacher, and began to produce no fewer than nine surviving offspring. Being something of an ideologist, with a wide circle of contacts and not completely at odds with the *l'ideal Français,* Hassler found himself being drawn ever deeper into the democratic processes of the new republic. Elected deputy governor of Aargau, he became one of the French-style attorney-generals of Switzerland. Politics and business began to dominate his life, which is fortunate, as the Swiss trignometrical survey was on indefinite hold. In 1803, the Swiss government was totally reformed, and Hassler, now 33, left Berne to return to his native Aarau. In the same year, the French administration decided to take command of the Swiss mapping, placing it under the direction of the members of the newly reformed and politically correct Académie de Science. At this point, or perhaps a little earlier, Johann Tralles chose to escape from the republican French and return to the Berlin Royal Academy, where he took up the prestigious post of professor of mathematics. He died in London in 1822.

With his old friend gone and the prospect of his beloved survey passing out of his hands, Hassler also decided he would leave to start a new life. France and America were still basking in the aftermath of victory from their joint triumph over the hated British at Yorktown. Prospects for new emigrants to the Americas looked good, and like the pilgrims of old, Hassler recruited some 120 Swiss pioneers to form a new American settlement. In 1804, by then comfortably wealthy, Hassler invested heavily in land speculation in South Carolina and in the vast expanse of Louisiana.

In Paris, the U.S. minister to France, Robert Livingston, was also frantically trying to acquire land in Louisiana, specifically near the vital gulf coast port of New Orleans. Louisiana, retroceded to France by the Spanish court in 1800, was a truly vast territory, which posed a serious threat to U.S. aspirations. Worse, its great seaport amid the swamps of the Mississippi delta controlled the passage of some 40 percent of America's exports. The United States desperately needed land or, better still, the entire city of New Orleans. But Paris was showing reluctance to do business. Just when the situation seemed hopeless, Livingston, through an admirable sleight of hand, won—simply by putting it about that the United States was considering a rapprochement with Great Britain. Such an idea was exactly what France feared—a reunited Anglo-Saxon empire.

Ferdinand Hassler (1770–1843). Courtesy NOAA Picture Archives.

This gross misrepresentation by the Americans induced Napoleon Bonaparte to react favorably toward the U.S. request; instead of granting its wishes, the emperor sold the United States his entire Louisiana territory, all 828,000 square miles of it, for a mere 60 million francs, or $27,267,622. Overnight, the land area of the United States more than doubled. But Napoleon was no fool; apart from the useful cash, "the sale assures forever the power of the United States, and I have given England a rival who, sooner or later, will humble her pride."

The Louisiana Purchase, as the deal became known, was 18 months old when Hassler and his Swiss pioneers set out for America. Departing from Amsterdam in the fall of 1805 aboard Hassler's chartered ship, the merchantman *Liberty,* the band of pioneers ventured into the stormy Atlantic. An apocryphal story is that, while on passage, the *Liberty's* captain was taken ill during a particularly foul storm, and Hassler was obliged to take over the navigation. He was appalled at the quality of the American coastal charts upon which he had to rely, and it was only through his skill that the ship was brought safely to port in the Delaware River in October.

This event, so the story has it, encouraged Hassler to think about the dire state of mapping in the United States. He determined to do something about it. Having survived an ocean passage of 2½ months, Hassler was devastated to learn that the land company in which he had invested heavily had failed: all his investments had been lost. With hardly any money and the well-being of a band of settlers and his own family to care for, Hassler was forced to sell most of his personal possessions, including much of his great library. Luckily for him, Philadelphia was the center of American science. Without further ado, he began to seek out contacts, intent on making friends among the members of the American Philosophical Society.

It was through his association with the society that he formed a friendship with John Vaughn, a wealthy wine merchant, and with Dr. Robert Patterson, scientist and director of the Philadelphia mint. Through Vaughn and Patterson, Hassler was introduced to the society's president, who at the time just happened to be Thomas Jefferson, then in his second term as president of the United States. This introduction was to serve Hassler well in the future.

In February 1807, Congress passed an important bill:

> That the President of the United States shall . . . cause a survey to be
> taken of the coasts of the United States, in which shall be designated
> the islands and shoals, with the roads or places of anchorage, within
> twenty leagues of any part of the shores of the United States; and also
> the respective courses and distances between the principal capes, or
> head lands, together with such other matters as he may deem proper
> for completing an accurate chart of every part of the coasts within
> the extent aforesaid.[6]

It seemed a perfect, almost too perfect, opportunity for Hassler. One has to suspect that he had a hand in preparing it.

Dr. Caspar Wistar of the Philosophical Society, a noted nationalist, also thought it a suspect coincidence and wrote to Jefferson expressing his concerns. Jefferson's response was sympathetic: "The foreigners who come to reside in this country, bring with them an almost universal expectation of office . . . the trusts of every country are safest in its native citizens." But the president concluded his reply on a more conciliatory note, observing "some employments . . . meritorious foreigners &c of peculiar qualifications may sometimes be introduced, such is the present case."[7]

For the sake of democracy, a notice was published, inviting scientists in America to present to the government their ideas for making the survey a reality. Hassler, of course, responded so quickly that it has to be supposed that he was on the inside track all the time. In all, twelve responses were received, including submissions by America's leading earth measurer, Andrew Ellicott (1754–1820), and James Madison. A committee of the American Philosophical Society headed by Hassler's good friend and patron Robert Patterson evaluated the proposals. Hassler was gratified, though not surprised, to learn that his proposal was the successful one. This outcome soured relations with Ellicott and was to lead to unfortunate consequences.

By this date Hassler had secured a professor's job at West Point Military College; in May he was presented with a sheaf of notes and sketches, along with $100, by Captain Meriwether Lewis, recently returned from the famous Lewis and Clark expedition to discover a route across the Louisiana Purchase to the Pacific Ocean. On President Jefferson's instructions, Hassler was to calculate the expedition's astronomical observations for latitude and longitude in order to provide an accurate framework for Lewis and Clark's map.

Throughout their expedition, Lewis and Clark had observed lunar distances for longitude at many locations along the route, but they did not compute the results. To waste time over such matters would have been futile on such an arduous and dangerous journey. Instead, Lewis's instructors, Andrew Ellicott and Robert Patterson, advised that only the observations should be recorded, leaving the computations to the experts to sort out when they returned. When Hassler examined the astronomical notes, he was perplexed. Unbeknown to him, the method of lunar distances used by Lewis, devised by Ellicott, and improved by Patterson dispensed with the need for the usual requirement to measure the altitude (angle above the horizon) of the moon and reference stars.[8]

In 1809, the professorship at West Point ended because of a lack of funding, but Hassler managed to secure a similar position at Union College. He was still trying to make sense of Lewis and Clark's figures in 1810 when he wrote to his friend and benefactor Robert Patterson describing his difficulties. The mystery is why couldn't, or wouldn't, Hassler do the arithmetic? Patterson had produced worked examples, which he must have discussed with Hassler, yet for some reason Hassler refused to accept the validity of the Patterson-Ellicott formula. To compound his troubles, progress on the proposed coast survey was at a standstill because of the difficult political situation that had recently developed with France and England over their attacks on American shipping and trade.

When the furor subsided, Jefferson's presidential successor, James Madison, resurrected the idea for the survey. In 1811, Congress voted $25,000 for purchasing equipment. Hassler was dispatched to Britain to find and acquire the necessary instruments. Within 8 months of his arrival in London with his wife and family, America declared war on Britain (the War of 1812), and once again the sea-lanes were closed. Fortunately, the British government maintained its traditional position that it "makes no wars on science," the same edict given by Benjamin Franklin to American privateers about his friend James Cook's expeditions.

In his quest for surveying instruments and professional advice among the London scientists and instrument makers, Hassler needed some willing contacts. Nevil Maskelyne would have been the logical choice, but the old astronomer had recently died. Instead, Hassler turned to the new astronomer royal, Dr. John Pond. A list of instruments was drawn up, including a theodolite of 24-inches in diameter, an astronomical telescope, two transit instruments, baseline measuring apparatus, barometers for heighting, thermometers, a balance, and an English brass standard. For a supplier of such a large set of scientific apparatus, Hassler turned to "that distinguished artist, Mr. Edward Troughton of London, agreeably to our united views, and with that interest for its success, which the great friendship with which he was pleased to favour me could alone inspire."[9]

The next 3 years were torture for Hassler and his family. Troughton proved to be as dilatory for Hassler as Ramsden had been for Roy. The family took a house next door to Troughton's workshop in hopes of speeding up the manufacture of the instruments. In November 1814, Hassler moved his family from smoky London to live with friends in Paris for reasons of health. He was back in London in February 1815, only to discover that the U.S. Treasury Department, suspicious of his politics and prolonged stay in Europe, had suspended his line of credit.

On the outbreak of the last installment of the Napoleonic Wars, Hassler rushed to Paris to rescue his family from the clutches of the French, incurring yet further delay. Returning to London, he was relieved to find that Troughton had at last completed a number of his instruments, only then to discover that the Treasury Department was bickering about the instrument maker's £3,700 bill. The bureaucrats in Washington even refused to pay Hassler's salary or his passage home.

Eventually, through the good offices of John Quincy Adams, then in London on a diplomatic mission of political reconciliation, Hassler was able to conclude his business with Troughton and return with his long-suffering family to their home in Philadelphia. On 5 January 1816, Hassler resubmitted his plan for the survey of the coasts to the secretary of the treasury, Andrew Dallas. Finally, on 3 August 1816, 5 years after being first offered the job, Ferdinand Hassler was appointed the first superintendent of the famous United States Coast Survey.

With the new post came new responsibilities. The Lewis and Clark calculations, so necessary for the map of Louisiana, which Hassler had steadfastly failed to solve, were finally laid aside "in despair." Just why such important observations were entrusted to one man or why the American Philosophical Society and the U.S. government were satisfied to let the matter stagnate when there were others who could have completed the calculations is strange. The most likely reason was probably Hassler's somewhat dogmatic personality. A vain and prudish man, he always wore woolen underwear, winter and summer, and imported Swiss wine and cheeses rather than buying American products. He even went to the extreme of purchasing Hungarian printing plates for his maps, and he hired German engravers rather than employ some of the excellent artists of Philadelphia.[10] These traits did not endear him to many in the American establishment and alienated him from others, but none could challenge his remarkable abilities or his contribution to American earth-measuring science.

As late as 1817, Thomas Jefferson still harbored hopes "to have made, at the public expence, the requisite calculations, to have the map corrected in its latitudes and longitudes, engraved and published on a proper scale."[11] But he was to remain disappointed. Lewis and Clark's long days and weary nights observing the heavens for science were completely wasted: their valuable measurements were assigned to oblivion.

# 23

## MEN WORTHY OF CONFIDENCE

THE FINAL SAGA IN THE STORY of the eighteenth-century earth measurers was written during its closing years. For as long as men had been measuring distances, the units they used were based on what they could physically manage. The span, hand, pace, foot, and inch had their equivalents all around the world.

Since the birth of civilization, humanity had striven to standardize its units, so necessary for trade, commerce, and taxation. England was one of the few European nations whose measures of length had some history of standardization. The story goes that it had been Edgar of Kent, a tenth-century Saxon king, who had made the first official "yard" from wood and kept it in the cathedral at Winchester. In 1196, during the reign of the absentee king, Richard I, Coeur de Lion, a court was held at Winchester, known as the Assize of Measures, that declared, "Throughout the realm there shall be the same yard of the same size and it should be of iron."[1] In the reign of Edward I, the yard was even better defined:

> The Iron Ulna [yard] of our sovereign Lord the King contains three
> feet and no more: the foot must be of twelve inches, measured by the
> correct measure of this kind of ulna: that is to say, one thirtysixth part
> of the said ulna makes one inch, neither more nor less. Further it is
> ordained that three grains of barley, dry and round makes one inch,
> twelve inches makes one foot; three feet makes an ulna: five and a half
> ulna makes one perch: forty perches in length and four perches in
> breadth makes one acre.[2]

In the Middle Ages, the perch, or rod as it is also known, an old Anglo-Saxon land measure, was used for the sake of conformity in measuring and marking out the land strips of the old English open field agricultural system, defined as the distance made by the left feet, toe to heel, of the first sixteen men to exit church after matins. In 1588, Queen Elizabeth had a new yard standard constructed, possibly from old Edgar's wooden stick, which remained the lawful English yard until 1824, when George IV had the modern yard defined.

In 1742, the Royal Society commissioned the instrument maker Jonathan Sisson to construct the first "imperial standard yard." Made of brass and actually engraved by his young apprentice, John Bird, this was the standard of length employed by Mason and Dixon, Nevil Maskelyne, William Roy and the early Ordnance Survey, and Rueben Burrow and his contemporaries in India. It was the also the standard that the Royal Society swapped with its colleagues at the Académie Royale; a copy was even taken by Hassler to America.

Until the dying years of the eighteenth century, despite many good intensions, there were no internationally agreed-upon common units or any direct physical relationship among units of distance, weight, and volume. Nowhere was this more evident than in France, where every province, every town, had its own standards of measure.

The dream for a universal unit was nearly as old as science itself, but it took a bloody revolution to make the dream a reality. The desire for a universal unit among France's academicians dated back to 1670, when Gabriel Mouton, a Lyonnais Jesuit priest, proposed a unit based on 1 minute of arc of longitude divided into ten parts. In 1671, Jean Picard had suggested a "universal foot," which he defined as one-third of the length of a pendulum beating the seconds. Fifty years later, Cassini proposed that a French foot (*pied*) should be defined as 1/6,000 of a minute of arc, based on his meridian arc observations of 1740.

In August 1789, as the first breeze of the French Revolution began to blow about the ears of Louis XVI, the beleaguered king signed a decree abolishing the royal monopoly on weights and measures in favor of a proposal to develop a unified system. The Chevalier Jean-Charles de Borda (of Borda circle fame) had recommended a new decimal measure, and the Académie Royale des Sciences was tasked with studying his ideas. In 1790, Charles Maurice de Talleyrand Perigord (1754–1838), bishop of Autun, presented a proposal to the National Assembly for standardizing the new unit based upon Picard's idea. The proposal was to define the unit as representing the length of a pendulum beating seconds, oscillating at the midlatitude point of 45° north. Talleyrand's proposal was passed to the Committee on Agriculture and Commerce and cleared to receive the assent of the doomed king.

Talleyrand was anxious for the new unit to have a truly international flavor, which meant having the British on board. From Paris on 29 March 1790, Talleyrand wrote to the British member of Parliament Sir John Riggs-Miller:

I understand that you have submitted for the consideration of the British Parliament, a valuable plan for the equalisation of measures: I have felt it my duty to make a like proposition to our National Assembly. It appears to me worthy of the present epoch that the two Nations should unite in their endeavour to establish an invariable measure and that they should address themselves to Nature for this important discovery. If you and I think alike on this subject, and that you axe of opinion that much general benefit may be derived from it, it is through you only that we can hope for its accomplishment; and I beg to recommend it to your consideration. Too long have Great Britain and France been at variance with each other, for empty honour or for guilty interests. It is time that two free Nations should unite their exertions for the promotion of a discovery that must be useful to mankind.[3]

Unfortunately, it seemed that Great Britain and France had indeed been too long at variance, and Parliament threw away the olive branch along with the chance for reconciliation in both measures and peace.

In March 1791, a committee was formed of the leading measurement experts of the day including de Borda, Antoine-Laurent Lavoisier, Jacques-Dominique Cassini, Louis de Lagrange, Pierre-Simon de Laplace, Monge, and Marie-Jean de Condorcet. These savants saw in the philosophy of new units "a means of spreading Enlightenment and fraternity among all people." Most of these men were of the upper classes, but they were adept and wise enough to embrace the spirit of the revolution in their report. The academicians proposed that the new Republic's new unit of length should be "for all people, for all time."[4] The recommendations were as revolutionary as the political revolution sweeping dangerously across the country.

The new unit of length would be "natural" and would be based on the size of the earth itself. De Borda suggested calling it simply the "meter" (from the Greek *metron,* measure). The commission agreed that the definition of the meter would be the unitary length of 1 ten-millionth part of the distance along the meridian passing through the Paris Observatory from the North Pole to the equator. In proposing this radical theology, the French academicians had a flying start, thanks to Cassini's French arc measure of 1740 and the results from their expeditions to Lapland and Peru. Nevertheless, if the wonder unit was to be a success, the scientists needed an even more precise determination for the size of the earth. A completely new meridian arc measurement was the only solution. The arc would be the longest ever undertaken and would extend either side of the 45th parallel, halfway up the world, extending from Dunkerque in the north of France to Barcelona in the south of Spain.

The bold proposal was presented by Comte de Condorcet to the French Parliament and was immediately approved. The work of the project was divided between Jean-Baptiste Delambre and Pierre-François Méchain, who would be responsible for the triangulation and the determination of the latitudes. Antoine Laurent Lavoisier (1743–1794), the father of French chemistry, and de Borda were to design the baseline measurement apparatus.

Also accepted was Talleyrand's proposal for a "natural" unit defined by the beats of a one-second pendulum in the same manner in which Picard had preserved the length of the Paris toise. This delicate work was assigned to de Borda and Cassini, who would be using the platinum pendulum of the Paris Observatory. De Borda, Condorcet, Lagrange, and Lavoisier were appointed as the project's superintendents, appointments that were approved by the ill-fated King Louis.

Whatever history may say about Louis XVI, he had a cool head: the very day before the besieged monarch attempted to escape from the revolution by fleeing to Varennes, he met with Cassini to discuss the project. When the philosopher had explained his plan, the king asked, "You are repeating the measurement of the meridian line which your father and your grandfather already made before you. Do you believe you can do better than they?" Cassini replied, "I would certainly not flatter myself to do better, had I not a great advantage over them. The instruments that my father and my grandfather used gave the measurement of the angles to an accuracy of only 15 seconds. Chevalier de Borda has invented one which will give me this measurement of the angles to the accuracy of one second, it will be all to my advantage!"[5]

The instrument of which Cassini spoke was de Borda's repeating circle that the physicist had designed in 1786 for the trigonometrical survey linking the observatories of Paris and Greenwich. When running the chain of triangles from Paris to the coast, Cassini and Legendre had measured their angles with large quadrants while Méchain did the same with the repeating circle. De Borda's circle had demonstrated overwhelmingly its superior performance over the traditional instruments and was thought by the French academicians to surpass even Ramsden's great theodolite. In this they were mistaken, for the repeating circle suffered from a fault that none then appreciated.

The circle's accuracy came from its ability to derive mechanically an angular mean from multiple observations, the argument being that if an angle was observed and repeated all the way around the instrument's circle, the average of these observations would yield a higher accuracy than would a single observation. The flaw in this argument was that it ignored the instrument's systematic error, the inherent error, albeit tiny, from which every instrument, even modern ones, suffer. However, it would be many years before this was recognized, and the de Borda circle, ever an endangered species, survived well into the 1830s before being wiped out by the dominant Theodolitus Rex. Etienne Lenoir, France's leading instrument maker, constructed the de Borda circle along with all the principal instruments for the great project.

The measurement of the meridian arc to define the length of the meter began on 25 June 1792. Jean-Baptiste Joseph Delambre (1749–1822) was given the section to triangulate from Dunkerque, on the north coast of France, to Rodez, a town about 50 miles north of the Pyrenees, a distance of some 470 miles (760 km). An eminent mathematician and astronomer, Delambre was born at Amiens and educated at the College de France. Charles Mason's friend Jérôme Lalande, recognizing young Delambre's remarkable abilities, personally instructed Delambre in his science and even entrusted him with his most complex astronomical calculations. In 1781, the académie gave Delambre the task of determining the orbits of Uranus, the planet recently discovered by Hershel in England. He also compiled tables for the sun and the motions of Jupiter and Saturn and refined the tables for the eclipses of Jupiter's moons. He became a full member of the académie in 1792.

Years later, in 1809, at the request of Sir Joseph Banks, president of the Royal Society, Delambre was asked to secure the release of the French-born James Smithson (1765–1829), then being held as a prisoner of war by the French army during the Napoleonic conflict. Smithson went on to amass a fortune and, in his will, bequeathed much of his estate to the United States of America to found "an establishment for the increase and diffusion of knowledge among men." In 1846, Congress used Smithson's extravagant gift to found the famous Smithsonian Institution.

The most difficult section of the meter's meridian arc, extending north from Barcelona across the Pyrenees to Rodez, a distance of some 210 miles (340 km), was given to Pierre François André Méchain (1744–1804). Born in Laon and the son of an architect, Méchain had studied physics and mathematics but, because of financial problems, was forced to abandon his studies without qualifying. Like Delambre, he was befriended by Lalande, who secured him a position as assistant hydrographer at the Depot of Maps and Charts in Versailles. In this post he became involved in the surveys of the coastline of France. As an astronomer, Méchain discovered many new nebulae and comets. In 1787 he was teamed with

Jean-Baptiste Joseph Delambre.
Courtesy Library of Congress.

Cassini and Legendre to measure the French triangles linking the Paris Observatory with Greenwich.

In 1794, while Delambre and Méchain were engaged in the tedious, dangerous business of triangulation, the ad hoc revolutionary Committee of Weights and Measures, which had replaced the disgraced royalist académie, was considering the other aspects of the great project. Unfortunately, one of the key members, the chemist Lavoisier, was missing from their company. He had been one of the king's so-called "tax-farmer generals," responsible for the collection of the royal tithe, and had been arrested by the Revolutionary Council.

De Borda, risking his own neck, begged for the chemist's release from prison. "The permanent attendance of Citizen Lavoisier," wrote de Borda, "is irreplaceable because of his particular talent for all that requires precision. It is vital that this citizen be returned to his important work that he always follows with as much zeal as activity."[6] The answer to his plea was not long in coming and was not particularly encouraging. A letter advised De Borda that Claude Antoine, prieur de la Côte d'Or, an ex-academician and member of the notorious Committee for Public Safety, had been appointed commissioner responsible for monitoring the project. He would be attending the next ad hoc meeting to gauge the situation for himself.

The conclusion the prieur de la Côte d'Or and his revolutionary comrades drew was that "any function can only be performed by men worthy of confidence by their republican virtues and for their hatred of kings. Consequently," declared the Committee for Public Safety, "de Borda, Lavoisier, Laplace, Coulomb, Brisson and Delambre are expelled from the ad hoc committee of Weights and Measures."[7] It was a serious reversal, with a most unwelcome, potentially frightening outcome. Fortunately, the substitute commissioners installed by the Committee for Public Safety were mostly from the disfavored academicians' own revolutionary student fanatics, who all turned out to be quite hopeless.

After some 16 months of this dangerous nonsense, following a change in the regime, five of the expelled scientists were reinstated to their old positions, their revolutionary stand-ins being packed off to jail. Unfortunately, Lavoisier, accused of harboring royalist sympathies, *il est condamné à mort puis guillotine*. The excuse given by the Committee for Public Safety for condemning the great scientist to such a terrible fate was that "the Republic does not need scientists" (a capricious lot, these enthusiastic revolutionaries).

Lavoisier's compatriot Condorcet was also arrested on similar charges but chose to commit suicide in his prison cell rather than face the stitch-dropping, jeering mob clustered around the blood-soaked scaffold in the Place de la Concorde. De Borda, with great courage, used the citizen revolutionaries' own tactics and vehemently turned on prieur de la Côte d'Or, blaming him for all their troubles and for causing such an unnecessary delay in the revolution's scientific labors. The Committee for Public Safety had their own man arrested, but he survived to die safely in 1832.

Free to renew the scientific struggle for the glory of "the citizens of the world," de Borda began work on the construction of the delicate and all-important baseline-measuring apparatus. Once again, de Borda's genius for the mechanics of science and his profound appreciation for accuracy were unleashed. The measuring rods he designed were 20 pieds long and made from two dissimilar metal strips: one of brass, the other of precious platinum. At one end they were fastened firmly together but were free to slide at the other. The lengths of the bimetal strips reacted differently to variations in temperature. By measuring the relative differences, using a specially engraved scale and microscope, a very accurate measure was possible. For calibration, the Règles de Borda, as these calibrated rods were known, were compared with the Toise du Pérou, the version of the Paris toise used on the French equatorial expedition to Peru[8] and preserved at the observatory.

Meanwhile, Delambre and Méchain were proceeding with observations of the angles in the long chain of triangles that spanned France and Spain. Their work was dogged by the revolutionary zeal of the parties that were competing for the political hearts and minds of the people of France, as well as the threat of a German invasion. Delambre started work around Paris before moving north toward Dunkerque, then south to rendezvous with Méchain at Rodez. He was continuously plagued by revolutionary guards, arrested on several occasions, harassed by suspicious country folk, and examined by distrustful town councils. Delambre's troubles, although significant, were only mild inconveniences compared with those of his colleague Méchain.

Méchain began in the south of France, progressing quickly into friendly Spain, running his giant triangles down from the precipitous Pyrenees to Fort de Jouy on the outskirts of Barcelona. Unfortunately for Méchain, Spain's friendship toward its northern neighbor was rapidly deteriorating. Spain shared the same fate as Britain with regard to its monarchy; whereas Britain had the benefit of a German prince, Spain had got itself a French Bourbon monarch, Charles IV. The Frankish king and the Spanish aristocracy feared that the revolutionary zeal of France might, if nothing was done, appeal to oppressed Spanish peasants, with the chance that Spain's aristocracy would end up in the same mess as their French cousins. To prevent this unwelcome invasion of their comfortable lives, Spain declared war on France.

In the confusion, Méchain found himself marooned in Barcelona, with few friends and little money. His woes were compounded when he sustained a very serious injury in a freak accident at a mill. His recovery was slow; in fact, he never fully recovered; nevertheless, as soon as he felt well enough, he forced himself to venture into the high mountains to continue observations as best he could. Among the high peaks of the Pyrenees he was subjected to climatic extremes and compelled to undertake dangerous climbs to the craggy summits.

Completing as much work as the land and political situation would allow, Méchain eventually escaped from Spain by way of Genoa, Italy, where he stayed until 1795 before returning to his family in Paris. His return to the city was an occasion for celebration (some believed he was dead); he was promptly elected into the reformed Académie des Sciences and the Bureau des Longitudes and installed as director of the Paris Observatory.

Since leaving Spain, Méchain had been tormented by a nagging doubt that he had made an error in measuring the latitude at Barcelona. One star in particular did not seem to give an answer consistent with its fellows, and this troubled him greatly. If his latitudes were wrong, the length of the meridian arc would be incorrect; then the new meter would be in error. As it happened, it did not matter, for something else far more insidious was lurking in the wings, waiting to be discovered.

In Paris, many of the new citizen politicians wanted to see the entire metric project fail because of the turmoil and unrest it would cause among the simple people of the country and the anger it would provoke from the market towns over the removal of their control of weights and measures. Yet, despite the revolution, its manifold and frequently bloody political reversals, the abolition of the "aristocratic" Académie Royale des Sciences in 1793, le Terreur, and the execution of King Louis XVI, the work on the meter went slowly forward.

Delambre and de Borda set out and measured the two baselines with the bimetallic rods, and the great triangulation scheme was tied to them. Additional latitudes were measured at three intermediate points along the vast baseline, which together with the latitudes derived for the two terminals, provided the scale for the meridian arc. Over the distance of some 680 miles (1,100 km) from north to south, the error between the two baselines as computed from the trigonometrical survey was just 1 foot (0.3 m).[9]

From their observations, Méchain and Delambre calculated that the difference in latitude between Dunkerque and Barcelona was 9°39'. Taking into account the oblateness of the earth, they deduced that the length of the meridian quadrant from pole to equator was 5,130,740 toises. On the face of it, their calculations produced a meter fractionally shorter than the provisional meter deduced from Bougeur's value for the size of the earth.

The observations and calculations of the commissioners were submitted for scrutiny and approval to an "international" convention of distinguished foreign academics. As Europe was still at war, only France's allies, Denmark, the Low Countries, Switzerland, and representatives of the Italian states and neutral governments were present. The delegates examined the academicians' records, checking the calculations until they were satisfied all was in order.

However, something was not quite right. The entire premise for the meter was that it would be exactly 1/10 million of the distance along the Paris meridian from the North Pole to the equator. As it was not only impractical but impossible to measure the entire distance from pole to equator, the scientists decided instead that all they had to do was measure a short section of the arc and use the shape of Bouguer's ellipsoidal figure for the earth as a template. The earth is an oblate spheroid or an ellipse of rotation; therefore, the section from pole to equator should be a quarter of this ellipse. The curve of an ellipse is an easy piece of mathematics and produces a smooth, predictable curve. Imagine, then, the scientists' horror when they discovered that the curve of their meridian arc was anything but smooth. It undulated up and down. The four baseline lengths measured between the five fixed latitude points did not produce the predicted elliptical section.

Gravity's subtle ways were at work. The meridian arc crossed the western heights of the Massif Centrale, the vast stony heart of France. The Massif, a rugged plateau of high-density granite rock and ancient, silent volcanoes, was effectively cut off from the rest of France; its inhabitants had their own history and their own language. Farther south, the triangles climbed over the towering peaks and plunging valleys of the Pyrenean ranges.

Across this complex terrain, the attraction of mountains was attractive indeed; Maskelyne's experimental observations on Schiehallion were producing some very significant deviations to the academicians' plumb lines. Without any knowledge of how to correct for the terrible havoc gravity was wreaking on their survey, the citizen academicians were forced to privately admit that their struggle to define a "natural" unit of length had been a folly. Mother Nature was not about to let her foolish children exploit her oblate body for the sake of commerce and the universality of weights and measures.

Science, however, had triumphed over adversity. The results for the meter may have been disappointing for those in the know, but the discoveries in the process had added yet more valuable information to the slowly filling storehouse of knowledge about the shape and size of the earth. In the final analysis, it didn't really matter that the meter was imperfect. The overall distance between the two extremities was taken as the defining distance, wrong as it may have been. But the whole enterprise had been a truly inspired idea, thwarted by gravity's pernicious idiosyncrasies. This is not the place to explore the subsequent history of the meter except to say that it did, in the end, triumph to become the universal measure, for all people, and for all time, preserved in a single bar of precious platinum. Even the so-called "imperial" units—feet and inches, pounds and ounces—still preserved as the people's units in the United States of America, have as their scientific, defining standard the imperfect meter.

Still agonizing over his possible error for the latitude at Barcelona, Méchain in 1803 was granted his request by Napoleon to return to Spain, ostensibly to extend the chain of triangles across the sea to the Balearic Islands of Ibiza, Majorca, and Minorca, but in fact to find his elusive "error." While working across the coastal swamplands of Valencia, extending the triangles with his de Borda circle, Méchain contracted malaria. He died, his son by his side, at Castillón de la Plana on 20 September 1804, not knowing that "his" error was something beyond his control.

# 24

# IRREGULARITIES WE
# HAVE DISCOVERED

PIERRE MÉCHAIN'S OPERATIONS AT FORMENTERA in the Balearic Islands were completed in 1806 by Jean-Baptiste Biot (1774–1862) of the Bureau des Longitudes, assisted by Dominique-Francois Arago (1786–1853), a young man who would one day discover new laws of magnetism and prove the wave theory of light. Then, assisted by the astronomer Claude-Louis Mathieu (1783–1875), Biot complete the French meridian arc observations at Bordeaux before traveling to its northern terminal amid the coastal sand dunes at Dunkerque. There, in the old bell tower of the church, the two scientists set up the delicate seconds pendulum that had been calibrated against the Paris standard regulator, and diligently recorded its rate and measured the miniscule change in its time-keeping. At Dunkerque, being a shade closer to the earth's center than Paris is, the pendulum swung a little quicker.

Biot, a strange, conceited sort of person, something of a dandy and royalist agitator in his youth, was born in Paris and educated first at the Collège de Louis le Grand, then, after the briefest spell in the army, at L'École Polytechnique. Elected to the chair of mathematical physics at the Collège de France in 1800, Biot was admitted to the Académie de France in 1803. In 1804, he joined the growing band of aerial adventurers when he ascended 13,000 feet (4,000 m) in a gas balloon to make scientific and atmospheric measurements.

Within months of the allied victory over Napoleon at Waterloo in 1815, French and British scientists renewed their customary, friendly exchange of letters. The following year, Dominique-Francois Arago wrote to William Mudge of the Board of Ordnance in London about an idea for extending the French meridian arc northward into Britain. So enthusiastic were both camps for a restitution of normal scientific relations that, in a short matter of time, the Bureau de Longitudes de Paris and the Board of Ordnance had agreed to the idea in principle and set about the business of designing the survey. The French wanted to extend the "European arc," as it was now being called, as far as Great Yarmouth

Jean-Baptiste Biot (1774–1862).
Courtesy Library of Congress.

on the Norfolk coast, but the British had a grander scheme in mind: Mudge proposed to carry the arc far to the north, into the remote Shetland Islands.

In May 1817, Jean-Baptiste Biot arrived in England to work with Mudge on the planning for the meridian arc observations. It was an opportunity for the two men to consider the latest findings and theories for "attractive matter" with colleagues at the Royal Society. Like Mudge, Biot had discovered that the more scientists examined the earth's shape and form "microscopically . . . the more irregularities we have discovered."[1]

And it was these strange, elusive irregularities that troubled Mudge so much, as they had the savants of France when they struggled with the measures for the meter. The results of Mudge's own meridian arc from Dunnose to Clifton seemed to suggest that the earth, at least in the vicinity of southern Britain, was prolate; even so, he was convinced that what he was seeing was the consequence of some sort of gravity anomaly and not a misshapen planet. In his opinion, the earth was likely to be of a general oblate form but that its "common meridians" were made up of "unequal and dissimilar curves."[2] In this he was essentially right, and he was describing what today we call the geoid, an irregular "geopotential" surface that is everywhere at right angles to the direction of gravity; the attraction of mountains is no respecter of man-made laws and conventions.

Despite their very different characters and imperfect knowledge of each other's language, a friendship of sorts developed between Mudge and Biot as they traveled north to Edinburgh with Captain Thomas aboard HMS *Investigator,* the ship allocated to the expedition. Unfortunately for the entire enterprise, Mudge was taken ill at Edinburgh and, despite the attendance of the city's famous doctors, was obliged to hand over the project's operations to his young and talented but overeager assistant, Captain Thomas Colby (1784–1852). Colby had joined the survey in 1802 and, within a short space of time, had effectively, though not officially, become Mudge's deputy director. Relationships between Colby, the determined military engineer, and Biot, the arrogant academic, were strained from the start; perhaps Biot didn't realize that Colby had been in virtual control of the board's survey operations for some time, or perhaps their difference in age and character was the problem. Whatever the cause, the personal difficulties between the two men were to plague the mission, to the detriment of both nations.

At Edinburgh Castle, northern home to the Board of Ordnance and its survey operations, the party of observers assembled in mid-June to discuss the program of work. By this time Biot had agreed, albeit reluctantly, that the meridian arc should terminate in the Shetlands. After protracted discussions, the accord reached was that Colby and James Gardner, the chief draftsman from the Tower of London, together with Dr. Olinthus Gregory of the University of Edinburgh, should make the British latitude observations with Ramsden's zenith sector. Simultaneously, Jean-Baptiste Biot, assisted by Mudge's son Richard, would observe at the same points with de Borda's repeating circle. Biot, being expert in gravity observations, was to have charge of the pendulum measurements.

The true ground distance from Dunkerque to the site of the Shetland observatory would be determined at a later date by Colby, who would extend the British triangulation scheme from the mainland across to the islands. It would appear from a letter written in late 1812 by Simon Woolcot, Dalby's replacement at the tower, that plans for a link were already advanced and that Colby had already visited the islands. The letter ends, "I wish you fine weather for your zenith sector observations that you may not be detained long in the inhospitable Isle of Shetland."[3]

Within days of the Anglo-French team's arrival in early August at Lerwick, the fishing capital of the lonely isles, what had augured so well in May had broken down into a factious standoff. The unpleasantness between Colby and Biot, whatever its cause, was contagious; soon the party was divided into two contending camps: "Dr Gregory attached himself to Captain Colby; whilst the latter disliked and distrusted Biot, he, Biot, detested Dr Gregory."[4] Gregory did not

Mudge's British arc and the Anglo-French arc.

much like Biot either: "I do not hesitate to say that I never met so strange a compound of vanity, impetuosity, fickleness, and natural partiality, as is exhibited in his character."

Colby and Gregory left for Balta Sound, where they set up Ramsden's zenith sector and observed the latitude. Biot and Richard Mudge left for the remote northern island of Unst, where they set up the de Borda repeating circle in a temporary observatory on the precipitous slopes of Saxa Vord. Thus a direct comparison between two famous geodetic instruments, so hoped for by Biot and Mudge, did not occur in the Shetlands or, on another occasion, at the church tower of Dunkerque. What began in high hopes as a collaborative effort between two great scientific nations ended in wounded pride and bruised egos. As an example of mending scientific and diplomatic fences after the disastrous Napoleonic conflict, the experiment was an abject failure. As a scientific experiment, the results from the 1,400-mile-long meridian arc that should have spanned Europe, from the sunny Balearic Islands to the chilly stormy shores of the Shetland Islands, were a chimera.

Disappointed but never downhearted, William Mudge recovered from his illness at Edinburgh and returned to London to continue his inspiring directorship of the Board of Ordnance's survey. His reputation as a mathematician and observer of the physical sciences grew to such an extent that he was appointed governor of the Royal Military College and as the public examiner of Addiscombe College, where the East India Company educated its engineers. Tragically, he died from exhaustion in 1820 at the age of 57.

The Europe that emerged from the ashes of 1815 was radically changed in every respect. The golden century, too, had passed away, with the seeds of the Great War of 1914–1918 sown on the battlefields of Waterloo. Belittled as "a nation of shopkeepers," Britain would become in the nineteenth century the greatest power the earth had ever known. Its immense empire would stretch across Canada, Africa, India, and the Far East and across the Caribbean and the Pacific to Australia and New Zealand—until a quarter of the world could claim to be citizens of the British Empire. This vast empire had to be measured and mapped; thousands of surveyors were sent forth to cover the globe with their triangles, baselines, and meridian arcs.

Four years after Burrow's tragic death at Buxor, the surveyor William Lambton (1753–1823) arrived in India. His passion was geodesy, the accepted term for the study of the shape and form of the earth. On 10 April 1802, Lambton began measuring his first baseline for a chain of triangles extending across the Mysore plateau to a base of verification near Bangalore. For the next 20 years, Lambton and his team of surveyors from the Trigonometrical Survey of India carried the triangulation network along the Malabar coast and down to Cape Comorin at the southern tip of India and across to the central city of Bangalore. In time, this scheme of triangles became known as the Great Arc of India.

In 1818, a young, confident surveyor named George Everest (1790–1866) joined Lambton's trigonometrical survey. Two years later, young Everest contracted fever and was sent to Cape Town for recuperative leave. For some light relief, Lambton suggested that Everest apply his mind to the mysterious problem of La Caillé's meridian arc of 1755 and its strange consequences for the shape of the earth. Inspecting the sites of the triangulation stations and viewing the grand landscape of the cape firsthand, Everest concluded that the problem was the old enemy, the attraction of mountains, or as he put it, "a gravitational discordance" caused by the considerable mass of Piquetberg and Table Mountain. In the event, the arc was totally remeasured in the 1840s by Sir Thomas MacLear, Her Majesty's astronomer at the cape. MacLear confirmed Everest's hypothesis when he discovered a deflection in the vertical of some 8 seconds of arc, caused by the attractions of the cape's mountainous terrain. There was, after all, no error in old La Caillé's measurements.

Isaac Dalby, the unsung hero of geodesy, retired from the Board of Ordnance in 1799 to become professor of mathematics at the Royal Military College in High Wycombe. Within a few years of his appointment, the college moved to Surrey, where today it is better known as Sandhurst Royal Military College. Isaac Dalby died at Farnborough, Surrey, in 1824 at the ripe old age of 80. Six years later, Major James Rennell, the father of modern geography and mapper of Bengal, also died, aged 87. Such was his fame and depth of learning and so acclaimed his accomplishments that he was buried in the nave of London's magnificent Westminster Abbey.

In 1798, at the age of 61, Henry Cavendish made a new determination of the density of the earth. It had been Cavendish who had supervised the measurements on Hounslow Heath, had challenged Maskelyne's preconceptions about "attraction," and had been indirectly responsible for encouraging the astronomer royal to conduct the Schiehallion experiment in the first place. The Honourable Henry Cavendish was probably the century's most ingenious scientist and must rank at the top of its most peculiar person's list. A millionaire, he was described by Biot as "the richest amongst the learned and the most learned of the rich." He was unbelievably shy, absent-minded, and terrified of women. He was at Cambridge with Maskelyne but left without qualifying and could only socialize with the closest of scientific friends. A "silent, solitary man, he had his magnificent library in London, four miles from his residence on Clapham Common, so that he might not encounter persons coming to consult it. His female domestics had orders to keep out of sight, on pain of dismissal. His dinner he ordered daily by a note placed on the hall table." Lord Brougham said of him that he "probably uttered fewer words in the course of his life than any man who ever lived to fourscore years, not at all excepting the monks of La Trappe."[5]

During the course of his bizarre life, Cavendish made many experiments in both physics and chemistry; the world-famous Cavendish Laboratory at Cambridge is named in his memory. Cavendish's own gravity experiment built on the pioneering work of his friend John Mitchell (1724–1793), fellow of Queens'

Henry Cavendish (1731–1810). Courtesy Library of Congress.

College, Cambridge, who had died before completing his research. Cavendish used Mitchell's torsion balance and its heavy lead balls to determine the value of the gravitational constant, $G$, which in turn allowed him to calculate the mass and density of the earth and so doing "weighed the world." Cavendish calculated that the mean density of the earth was 5.448 times that of water, 20 percent greater than Maskelyne and Hutton's value.

Although his results were significantly different from those obtained by Maskelyne, Cavendish was extremely tactful toward his old friend over the difference: "According to the experiments of Dr. Maskelyne, on the attraction of the hill Schehallien, the density of the Earth is 4½ times that of water; which differs rather more from the preceding determination than I should have expected. But I forebear entering into any consideration of which determination is most to be depended on, till I have examined more carefully how much the preceding determination is affected by irregularities whose quantity I cannot measure."[6] Had he gone on to explore those irregularities then, without doubt Cavendish would have come very close to unraveling the mystery of why mountains, as Maskelyne's and Bougeur's results suggested, do not live up to their attractive expectations. Henry Cavendish died alone at his house on Clapham Common in 1810.

In 1792, in the midst of revolution and growing hostility toward Britain, Jérôme Lalande in Paris published the third edition of his *Astronomie*. The new edition contained an appendix entitled "Tables astronomiques calculeés sur les observations les plus nouvelles." Lalande's coauthors were Jean-Baptiste Delambre and Charles Mason of the Royal Greenwich Observatory. Since turning down the offer to conduct the Schiehallion experiment, Mason had been working for the Board of Longitude, correcting and improving Mayer's lunar tables that complemented Maskelyne's annual *Nautical Almanac*. He had been contemplating a return to America for many years. As soon as the Revolutionary War was over and relations with the new United States improved, he decided the time was right.

His decision to emigrate was probably colored by the depression that was then devastating the woolen industries of his native Gloucestershire. So harsh were conditions that, in the years between 1783 and 1790, some 40 percent of the county's population made the same decision as Mason. Although still on good terms with Nevil Maskelyne (Mason's *Mayer's Tables of the Moon from the year 1780* contained a warm note of indebtedness from the astronomer royal), Mason's status in the scientific hierarchy was on the wane. He had applied to the Board of Longitude for the £5,000 longitude prize under the revised terms of the 1774 act but had received only £1,317, which Lalande said fell well short of what Mason was rightly due. The great catalogue of nautical stars that he diligently compiled from the observations of his old master, James Bradley, was included in the *Nautical Almanac* for 1773 and his *Places of the Moon* appeared in the 1774 *Nautical Almanac,* remaining in print until well into the next century.

In February 1786, Mason's father-in-law, Robert Williams, sold his house in Tetbury with the intention of joining his daughter and her husband on a new life in Philadelphia. Sadly, the old schoolmaster died before the arrangements were completed, but the proceeds of his estate, together with the award from the Board of Longitude, were enough for the family to fund their move. The packet boat bearing Charles and Mary Mason and their eight children arrived in Philadelphia on 15 September.

The passage had been rough, and Mason's illness, exacerbated by his chronic seasickness, grew worse. By the time he set foot in America, he was extremely ill. On 27 September he wrote "from the Sign of the George" on Second Street to Benjamin Franklin, pleading for help for his family. Knowing that his days were ending, Mason consigned his scientific papers to his old friend the Reverend John Ewing, provost of the University of Pennsylvania. Charles Mason passed away on 25 October, exactly 26 years to the day that he and Jeremiah Dixon first signed on with the Royal Society. He was buried by his family and friends in the graveyard of Christ Church, Philadelphia.

On 1 December 1792, Mary Mason petitioned the Board of Longitude for a review of her late husband's claim to the longitude prize. Sir Joseph Banks was in the chair when the petition was laid before the board. On reflection, Banks decided to refuse the claim but, by way of compensation, offered Mary a meager £100 "for the Papers that she has . . . and in case of her refusal that they be returned to her whenever she chuses to demand them."[7] Supremely self-confident in his own genius, immensely wealthy, and a beneficial patron to many, Banks remained coldly uncharitable toward the astronomers. This ungrateful act on the part of the Board of Longitude is explained in part by the monopoly of the sciences by Bank's party and its advocacy of the naturalists over the mathematicians. Nevertheless, his decision to refuse a petition for what was undeniably due to a great scientist reflects poorly on his character at the time.

Never doubting the significance of his "important observations," the old Astronomer Royal Nevil Maskelyne had died at his post in the Royal Observatory in 1811, aged 80, unaware that both his and Cavendish's figures for the earth's density were essentially correct. Maskelyne had outlived most of his contemporaries. The experiment he and Reuben Burrow had conducted on the misty Scottish peak had transformed humankind's view of the earth forever and confirmed that the nebulous attraction of mountains really did exist. Thereafter, all astronomers, surveyors, and geodesists could, if not correct for the effect, at least be aware of the precautions necessary to obviate the worst of its consequences. The man who did the calculations, Charles Hutton, lived to see the national mapping organization he had helped nurture grow and flourish.

In the year Maskelyne died, his friend John Playfair, professor of natural philosophy in Edinburgh, performed a geological survey of Schiehallion to determine the nature and density of its rocks. As a result he revised upward Charles Hutton's value for the earth's mean density to between 4.56 and 4.87 that of water. Hutton's own conclusion was that the earth's density was more likely to be 4.95 that of water. Hutton, by then an old man, in a prickly paper to the Royal Society in 1821, reminisced how "by some strong mistake, or perversion, for many years, it was customary among certain persons [Nevil Maskelyne], to withhold the mention of my name with regard to the great share that I had in experiment on Schiehallien." Charles Hutton died peacefully in 1823, at the age of 86.

In the years following his death in 1810, Cavendish's hypothesis that the ocean produced a deficit of attraction was reexamined, as it suggested that the earth's interior was perhaps a more complex body than hitherto thought. Could it be that deep within the earth, great concentrations of mass, the remnants of prehistoric mountains, were producing the same effect as Schiehallion and the Massif Centrale? The last chapter in the story of the attraction of mountains was about to be written with a discovery of a new and peculiar sort.

In 1829, George Everest, superintendent of the Indian Trigonometrical Survey, was in England on recuperative leave. Everest, the man who never actually saw the mountain named in his honor, had succeeded William Lambton, who had died tragically beside the road at Hinjunghat in 1823. On a number of occasions Everest and Thomas Colby, director general of the Board of Ordnance's survey, met to discuss their work and compare notes on the peculiar results that the attraction of mountains was giving. During their discussions, Everest offered Colby a trade—all his data on the Great Arc in exchange for Mudge's meridian arc work.[8]

The next year saw the publication of two new figures for the shape of the earth. The astronomer and mathematician George Biddle Airy (1801–1892) published his parameters first, which were derived from meridian arcs taken at different places around the globe. In Airy's time, he had the choice of some fourteen meridian arcs. He knew that the "greatest discordances" occurred near mountains, so he purposely rejected those arcs, such as the French arc in Peru, from his calculations. This left eight arcs with which to work: Mason and Dixon's American arc; Mudge's British arc, Gauss's Hanoverian arc, and two Russian arcs; two long French arcs; and, of course, the meridian arc information Everest had given Colby. The figure he calculated, Airy's Spheroid, became the official spheroid of Britain and its national mapping organization, the Ordnance Survey. Unfortunately, he was unable to incorporate into his calculation the European arc, which, had discord not ensued, would have greatly improved the final result.

Everest also published his version of a figure of the earth. However, it was derived principally from his observations of the Great Arc of India, supplemented with the arcs of France. Everest's spheroid, as it became known, remains to this day the official spheroid for India; why Colby's information was never incorporated into an "imperial" spheroid is a moot point. Returning to India in 1830 as surveyor general, Everest's stern but brilliant direction of the survey of India flourished, and the surveyors' measurements became ever more precise and plentiful. What had begun as the illusive attraction of mountains now became "the deflection of the vertical" as the Great Arc moved northward across the plains of India toward the massifs of the Himalayan ranges, where gravity's deviant effects made themselves felt.

But even where there were no mountains, the surveyors' measurements were giving some odd results. Just as Mudge had found a deviation in the vertical of his zenith sector amid the gentle hills of England, so Lambton had, around Bangalore, also "discovered . . . an agent unthought of in former days, viz a disturbing force occasioned by the diversity in the density of strata under the surface." Across the world, similar effects were being noticed: it seemed as if other forces besides mountains were affecting the surveyors' plumb lines. To make matters worse, or at least more complicated, the deviations that the mass of the Himalayan Mountains, according to the scientists' calculations, echoed the findings of Bouguer on Chimborazo 100 years earlier—were not living up to their Newtonian expectations.

The curious deficiencies in the observations in India were given to John Pratt, archdeacon of Calcutta, a brilliant mathematician and geographer, to ponder. His conclusion, published in 1855, was that the earth's crust possessed an undersurface that appeared to follow the spheroidal shape of the earth. Reflecting in part the thoughts of Cavendish, Pratt's inference was that the earth's crust, instead of being of uniform density, varied from high density beneath the deep oceans to low density in the high mountains. The observations from the Great Arc appeared to match the theory.

In 1835, George Airy succeeded John Pond as the seventh astronomer royal. A self-critical, rigid disciplinarian and oftimes sarcastic, Airy was talented but, as he himself noted, not brilliant. He was driven by his science, and his range of interests extended to the deviant effects of gravity. In 1854, he conducted an experiment, using a pendulum, to measure the difference in the strength of gravity at the top and at the bottom of a deep mineshaft at the Harton mine in South Shields. From the vertical separation and the variation in swing rate, he was able to calculate another new density for the earth. Published in 1855, this experiment, together with the odd results then arriving in England of the lower than expected attractive effects of the Himalayas, suggested to Airy that mountains had deep "roots" that, somehow or other, effected a deficit of attraction, just as did the deep oceans in Cavendish's hypothesis.

In contrast to Pratt's version of the theory, Airy concluded that the materials of the earth's crust were uniformly of a relatively lower density than a deeper and denser underlying region. Whereas Pratt thought of mountains as surficial artifacts, in Airy's view the mountains represented thicker sections of crust that plunged as deep within the earth as they soared high above it, so that they "floated," much as an iceberg floats, in perfect equilibrium upon the denser materials beneath. Under the deep ocean, the Airy model allowed for a thin crust and correspondingly deep water. Airy's thoughts also matched the data coming from India.

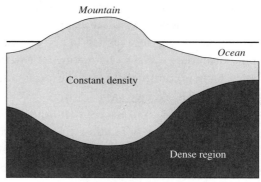

**Airy isostatic model**

Isostacy—Pratt and Airy's
hypotheses.

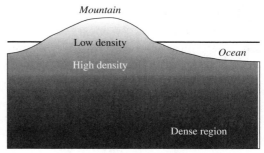

**Pratt isostatic model**

In 1889, the theory pioneered by Pratt and Airy became generally known as "isostacy," and the battle for and against its strange effects as being the real interpretation of what was going on beneath the surface of the earth raged throughout the nineteenth century and well into the twentieth. In 1926, the arguments were still raging, as Colonel Close, director general of the Ordnance Survey, recorded that "finality in the discussion of this matter of isostatic compensation has not yet been reached, though much progress has been made."[9]

With a vast body of observational evidence supported by conclusive proof from seismic measurements in the mid-twentieth century and later from satellite observations, isostacy was finally accepted as the fundamental principle and root cause for the attraction of mountains, the curious effect first imagined by Newton, tested by Bouguer in the high Andes, and proved by Maskelyne on Schiehallion.

The precision mapping begun by the Cassini family in France, Roy in Scotland, Rennell in India, and Hassler in the United States, placed within a firm foundation of knowledge about the shape and size of the earth, set the cartographic tenor for the colossal mapping programs of the nineteenth century. Rennell, accepting the weakness in his "astronomical" methods for locating his individual route surveys, championed the wisdom and expediency of triangulation for map control that led to the formation of the world's largest survey organization. The chance *mémoire* from Cassini, plus a little economy with the truth by Joseph Banks, ensured that Britain got its trigonometrical survey and Ramsden his chance of building the grandfather of all modern surveying instruments. For a century and a half, the triangulation principles pioneered by old Snellius were the primary means of earth measuring. In the 1950s a new dimension for triangulation was added with the introduction of accurate long-distance electronic measuring devices that gave the method a renewed, but transitory, lease of life.

Technology moves on, and the forlorn "trig pillars," those idiosyncratic monuments that mark the apexes of the earth measurers' triangles, are themselves becoming redundant as surveyors and geodesists forsake earthly measures and once again look to the skies for their position. Instead of quadrant and chronometer, they have a small box of tricks that uses the U.S. global positioning system (GPS) to tell them precisely where they are in a matter of minutes. What once took years can now be done in days. The era of the lonely surveyor, computing the day's work on a camp table by the light of a flickering hurricane lamp, has long gone. The field crafts handed down through the generations, the tricks of the trade that kept them alive and on schedule, are passing from memory.

Yet, although the science may have changed, the work goes on: even the high-flying satellites of the GPS cannot escape the attraction of mountains.

# Appendix: Explanations and Definitions

The following explanations are brief descriptions of what can be very complex subjects. These notes do not pretend to be anything more than readers' aids for appreciating some of the finer points of the story. For a more complete understanding, the reader is referred to the many excellent and popular books that are available, particularly on astronomy. Beyond these works is a wealth of technical literature that lends itself more to the student of geodesy, surveying, earth physics, geophysics, and astronomy.

## TIME

The nature of time is one of the more difficult concepts to grasp and, today, is a very complicated business. With the advent of worldwide coverage by GPS, precise time is available to everyone who has a suitable receiver. But GPS time is yet another example of a coordinated time frame that is independent of natural forces and is regulated by the world's most advanced atomic clocks. Fortunately, for this story we need only be concerned with solar, sidereal, and mean time.

### Solar Time

When the sun reaches its highest point in the sky and lies on the meridian, it is local noon for the observer. The interval between two consecutive "noons" is an apparent solar day. This is the time shown by a common sundial. The length of a solar day varies through the year because the earth's orbit around the sun is elliptical and not circular (see Kepler's laws of planetary motion). Because the earth's axis is tilted by 23.5 degrees to the plane of its orbit around the sun, in the northern hemisphere the planet's attitude toward the sun makes apparent solar days shorter between June and December and longer between December and June (vice versa in the southern hemisphere). In a solar year, the sun, as viewed from the earth, rotates 365.24 times around the earth.

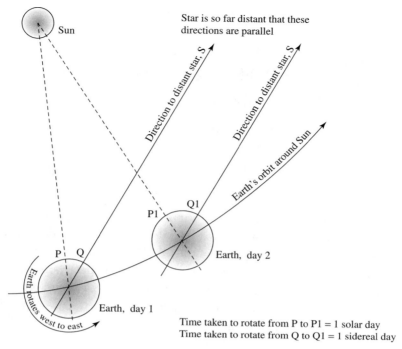

Star is so far distant that these
directions are parallel

Direction to distant star, S

Direction to distant star, S

Earth's orbit around Sun

Sun

Q1

P1

Earth, day 2

P    Q

Earth, day 1

Earth rotates west to east

Time taken to rotate from P to P1 = 1 solar day
Time taken to rotate from Q to Q1 = 1 sidereal day

Solar time and sidereal time.

### Sidereal Time

Sidereal time, or star time, is the time as reckoned by observing the stars. A sidereal day is the period elapsing between successive transits across the observer's meridian of the First Point of Aries or, indeed, any star. If, for example, the passage of a star across the meridian occurs at 15 hours and 20 minutes sidereal time, it will appear at exactly the same sidereal time the next day. In a sidereal year, the stars, as viewed from the earth, revolve 366.24 times around the earth, that is, one revolution more than in a solar year.

### Mean Time

Mean time, or mean solar time (also known as clock time), is based on an imaginary sun that travels at a constant velocity around the celestial equator (i.e., in a circular orbit). A mean time day is a constant 24-hours long and, being constant, can be replicated mechanically. The difference between apparent solar time and mean solar time is called the Equation of Time and can vary by up to 16 minutes through the year. Mean time is today regulated by the International Bureau of Time, and the world is divided up into time zones based (more or less) on Greenwich. This time system is called Universal Time and is the time shown by your watch.

## ASTRONOMICAL TERMS

### Kepler's Laws of Planetary Motion

Johannes Kepler first described planetary orbits in mathematical terms in his laws of planetary motion. He calculated that the orbits of the planets in the solar system are elliptical and not, as the Greeks thought, a complex mix of circular and epicyclic motions. An ellipse is a special form of circle with two centers, or foci. According to Kepler, the orbit of a planet is an ellipse with the sun at one of these foci. According to his laws, a planet travels around this elliptical orbit faster when it is nearest the sun and slower when it is farthest away.

### Orbit

The orbit of a planet, such as earth, is the path that it follows around its primary under the influence of gravity, in our case, the star we know as the sun. Modern relativistic physics offers a far more complex explanation.

### Celestial Sphere

The celestial sphere is an imaginary sphere of infinite radius with the earth at its center. The concept is Greek, and each planet is seen, including the sun, as rotating on its own crystal sphere, with the stars sprinkled across their own sphere. The whole wonderful cosmic machine is driven by yet another force called the *primum mobile.* Although the concept is wrong, viewed from the earth, it does appear as if the planets, sun, and stars wheel around the earth from east to west. For convenience of visualization and even for calculation, surveyors and positional astronomers continue to use the concept of the celestial sphere.

### Celestial Equator

The celestial equator is an extension of the earth's equator projected out into space until it meets the celestial sphere. It is the datum line from where declination is counted (see right ascension and declination).

### Ecliptic

As seen from the earth, the path that the sun follows against the fixed background of the stars in its annual voyage around the celestial sphere is called the ecliptic and is inclined to the celestial equator.

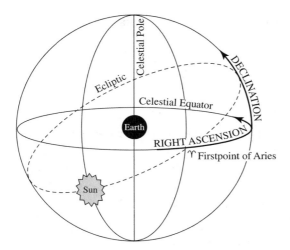

The celestial sphere.

### Equinox

As the sun moves around the ecliptic, it crosses the celestial equator twice, once going northward on or about March 22 and then again going southward on or about September 22. The springtime crossing point is called the First Point of Aries or Vernal Equinox, and the autumn or fall crossing is called the Autumnal Equinox. At the precise moment of crossing, the lengths of night and of day are equal, hence equinox. The First Point of Aries, symbolized by the zodiac sign for Aries, the Ram, ♈, is the point in space from where right Ascension is counted (see right ascension and declination). At the present time, the Vernal Equinox occurs in the constellation Pisces. When this enigmatic point was first identified over 2,200 years ago, the event occurred in the constellation Aries, hence First Point of Aries. In 600 years' time it will be in Aquarius. The cycle of the Vernal Equinox around the celestial sphere is known as the precession of the equinoxes, and it takes 18,000 years to complete one circuit.

### Right Ascension and Declination

The celestial positions of the stars, planets, and even the sun are described in a system similar to latitude and longitude. Right ascension (RA) is the astronomical equivalent of longitude and is measured from the First Point of Aries (the Vernal Equinox point). Measured in hours, minutes, and seconds, where 1 hour in time is equal to 15° in angle, RA is directly associated with sidereal time and can be used to determine the time a star will reach a certain point in the sky, for example, crossing the meridian.

Declination is the celestial equivalent of latitude and is measured in degrees, minutes, and seconds (of arc) north or south from the celestial equator. Every

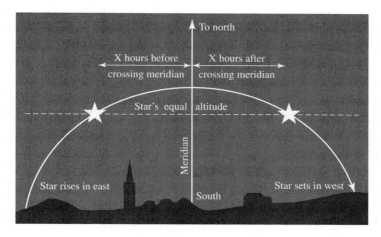

Finding the meridian by equal altitudes.

celestial object has an associated RA and declination, so if a star has an RA of 4 hours, 10 minutes, and 25 seconds and is on the meridian, a star with an RA of 14 hours, 20 minutes, and 50 seconds will be on the meridian 10 hours, 10 minutes, and 25 seconds of sidereal time later.

### Meridian

The local meridian is an imaginary line passing through the observer from the north celestial pole and through the zenith to the south celestial pole and, therefore, is a great circle (a circle that passes through the center of the earth) perpendicular to the horizon. The sun, planets, and stars cross the local meridian from east to west once a day as the earth rotates from west to east once a day. The RA of a celestial body as it passes the meridian is synonymous with local sidereal time.

The Greenwich meridian, passing through the old observatory in Greenwich Park, London, is a special meridian in that it is the origin for all longitude measurements. Before it was internationally accepted, other nations had their own zero meridians, for example, the one that passed through the Paris Observatory or even the one that passed through the White House in Washington, D.C.

## GEOGRAPHICAL TERMS

### Circular Measure

For more than 5,000 years, the circle has been divided into units. The most familiar units are degrees (°), minutes ('), and seconds ("). The notation is Mesopo-

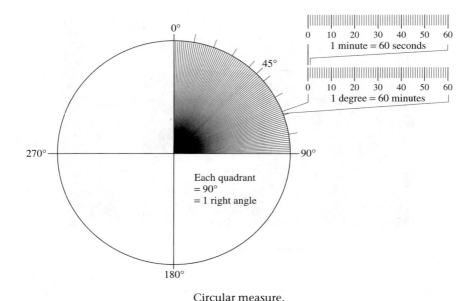

Circular measure.

tamian, whose people used the sexagisamal system, and is thus very ancient. The circle is divided into 360°. Each degree is divided into 60′ and each minute into 60″, thus the circle contains 360 × 60 × 60 seconds, or 1,296,000 seconds, in a whole circle. To differentiate angular minutes and seconds from those of time, it is common practice to add the term "arc," for example, 10 arc seconds. Common practice these days is to count the degrees clockwise from 0. In the eighteenth and nineteenth centuries, it was more common to measure horizontal angles from north or south on the compass dial. For example, 45° could be expressed as north, 45° east; or 200° as south, 20° west. Navigators and shipmasters counted the compass in an even more complicated system.

### Latitude and Longitude

A position on the earth can be precisely expressed in terms of its latitude and longitude. Latitude is expressed in degrees north or south of the equator, and there are three definitions:

*Geographic latitude:* the latitude north or south of the earth's equatorial plane measured from the center of the earth. It cannot be observed directly and has little relevance other than for small-scale atlas maps.

*Geodetic latitude:* the latitude north or south of the equatorial plane where it is intersected by a line normal, or perpendicular, to the spheroid. Geodetic lati-

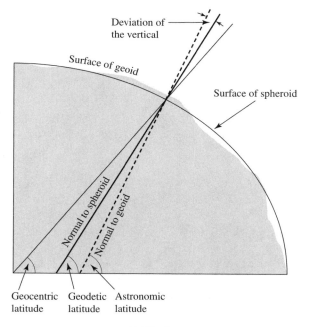

Latitude.

tude is therefore dictated by the spheroid in use, for example, Clark 1866, used in the United States. Geodetic latitude is the latitude used by surveyors but is not directly observable, because it requires knowledge of the deviation of the vertical, in other words, the attraction of mountains.

*Astronomical latitude:* the latitude defined by the direction of gravity at the observer's position. Astronomical latitude is the only latitude that can actually be observed by taking measurements from the stars.

Longitude is, these days, expressed in degrees east or west from Greenwich. In the eighteenth and early nineteenth centuries, it was more common to use hours, minutes, and seconds on the basis of 1 hour = 15°. Like latitude, longitude measured by astronomical means, for example, lunar distances, is affected by the direction of gravity and has to be adjusted. Harrison's chronometer (and the precision timekeepers of today) did not suffer from this difficulty.

### Figure of the Earth

In its earliest form, "figure of the earth" referred to its apparent "spherical" dimensions. More commonly it now refers to the parameters of the adopted reference spheroid.

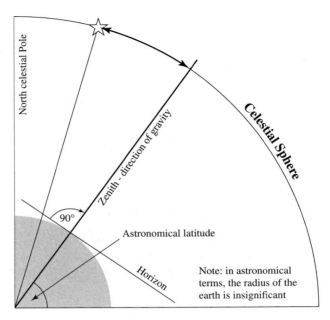

Finding the latitude by the zenith distances of stars.

*Spheroid*

A spheroid (or ellipsoid) is an ellipse of rotation whose dimensions are expressed in feet or meters in terms of its semimajor ($a$) axis and semiminor ($b$) axis. The ratio $(a-b)/a$ is the so-called compression or flattening. It is common practice to define a spheroid by its $a$ axis and flattening (or, if being particularly finicky, the eccentricity coefficient $e^2$). For national mapping purposes, a spheroid is chosen (or developed) that best fits the particular nation's section of the geoid and is known as a reference spheroid. For worldwide mapping or international projects, a reference spheroid is chosen that is the best fit for the entire earth, such as WGS84 (World Geodetic System 1984), developed from satellite observations specifically for the GPS navigation satellite network.

*Geoid*

The geoid is the "true" shape of the earth and is an imaginary surface that is everywhere perpendicular to the direction of gravity (an equipotential surface). It also approximates global mean sea level. If frictionless canals linked the seas and oceans of the world together, the resulting water level would form the geoid.

## Projection

A projection is a mathematical method of presenting the curved surface of the earth onto the flat plane of a piece of paper. The type of projection chosen has to satisfy the purpose of the map. For example, a political map might choose an equal area projection to show the counties at the true size.

## Tropics

The latitude 23.5° north and south marks the limits of the sun's twice-yearly journey across the equator. The northern limit is the tropic of Cancer, which the sun reaches at the summer solstice on or about June 22, and the southern limit is the tropic of Capricorn, which the sun reaches on or about December 22. The sun crosses the equator twice, once at the vernal and once at the autumnal equinox.

## Spherical Excess

The sum of the angles in a plane triangle, one drawn on a piece of paper and familiar from our schooldays, is 180°. However, the sum of the interior angles of a spherical triangle, a figure measured on the earth's curved surface, is always greater than this. William Roy developed a simple theorem for handling the spherical excess of his triangles, but it was the French mathematician LaGrange who developed the theorem used today. It requires a spherical triangle with an area of about 75 square miles to produce a spherical excess of 1 second of arc.

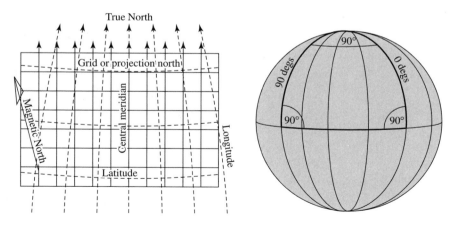

Spherical excess and north.

## North

For all practical purposes, there are three norths:

*True north:* also known as geographical north, it is the point where the earth's axis penetrates the planet's surface at the North Pole. The South Pole is the other end of the earth's axis. The North Pole and South Pole are, essentially, fixed points.

*Magnetic north:* the earth acts like an enormous bar magnet. The common magnetic compass points toward this bar magnet's "north pole." Magnetic north moves slowly and continuously and, therefore, is not a fixed geographical point (currently sliding across Ellis Island toward Russia). The angular difference between magnetic north and true north is known as the variation of the compass.

*Grid (or projection) north:* grid north is the common grid of a projection; for example, Britain's national grid or that of the United States has its central meridian (typically) aligned toward true north. Only along this central meridian are grid north and true north coincident. Elsewhere, grid north deviates from true north by an amount that increases the further one moves east or west.

# Notes

PREFACE

1. Frederich Engels, *Herrn Eugen Dührings Umwälzung der Wissenschaft* (Berlin, 1948), 321.

2. H. W. Brands, *The First American: The Life and Times of Benjamin Franklin* (New York: Anchor Books, 2002), 373.

3. Johnathan Swift, *On Poetry: A Rhapsody* (Dublin, 1733).

CHAPTER 1

1. J. M. Cohen, trans. *Christopher Columbus: The Four Voyages* (London: Penguin Books, 1969), 39.

2. Ibid., 43.

CHAPTER 2

1. Dieter Lelgemann, *On the Ancient Determination of the Meridian Length by Eratosthenes of Kyrene.* Presentation at History of Surveying and Measurement WSHS1, Federation International de Geometre Working Week, Athens, 2004.

2. Martin Gorst, *Measuring Eternity* (New York: Broadway Books, 2001), 39.

3. Nicholas Crane, *Mercator: The Man Who Mapped the Planet* (London: Weidenfeld & Nicolson, 2002), 154–160.

4. Charles Moreland and David Bannister, *Antique Maps—A Collector's Handbook* (London, 2000). Available at http://www.antiquemaps.co.uk/book/index.html

5. Derek Howse, *The Quest for Longitude.* The proceedings of the Longitude Symposium, Harvard University, Cambridge, Mass., Nov. 4–6, 1993, ed. W. J. H. Andrewes (Cambridge, Mass.: Collection of Historical Scientific Instruments, Harvard University, 1995), 150–151.

CHAPTER 3

1. P. Costabel and M. Martinet, "Morin, Quelques savants et amateurs de science au XVIIe siècle" (Paris, 1986).

2. Ibid.

3. Derek Howse, *The Quest for Longitude:* The proceedings of the Longitude Symposium, Harvard University, Cambridge, Mass., Nov. 4–6, 1993, ed. W. J. H. Andrewes (Cambridge, Mass.: Collection of Historical Scientific Instruments, Harvard University, 1995), 150–151.

4. F. Marguet, "Histoire de la longitude à la mer au XVIII$^c$ siècle" (Paris, 1917), 7.

5. James Boswell, *The Life of Samuel Johnson* (London, David Campbell [1791], 1992), 570.

6. J. J. O'Connor and E. F. Robertson, *Longitude and the Académie Royale* (St. Andrews, 1997). Accessed at http://www-history.mcs.st-andrews.ac.uk/HistTopics/Longitude1.html

7. Norman J. Thrower, *The Quest for Longitude.* The proceedings of the Longitude Symposium, Harvard University, Cambridge, Mass., Nov. 4–6, 1993, ed. W. J. H. Andrewes (Cambridge, Mass.: Collection of Historical Scientific Instruments, Harvard University, 1995), 55–56.

8. James R. Smith, *The Meridian Arc Measure in Peru, 1735–1745. Surveying and Mapping the Americas* (Washington D.C.: Federation International Geometre Convention, 2002).

CHAPTER 4

1. Gonzalez Antonio Moren and Pesar la Tierra, *Test y Origen de un Anacronismo* (Madrid: Departamento de Didáctica de las Ciencias Experimentales, 2000).

2. J. Kakkuri, T. J. Kukkamäki, J.-J. Levallois, and H. Moritz, *Le 250$^e$ anniversaire de la mesure de l'arc du meridien en Lapone* (Helsinki; Suomen Geodeettisen Laitoksen Julkaisuja [Finnish Geodetic Institute], 1986).

3. Pierre de Maupertuis, *Relation du voyage fait par ordre du Roi au cercle polaire pour déterminer la figure de la terre* (Paris: Mémoires de l'Academie Royale des Sciences, 1737, as quoted ibid).

4. James Boswell, *The Life of Samuel Johnson* (London: David Campbell Publishers [1791] 1992), 349.

5. Ibid., 349.

CHAPTER 5

1. James R. Smith, *The Meridian Arc Measure in Peru, 1735–1745, Surveying and Mapping the Americas* (Washington D.C.: Federation International Geometre Convention, 2002).

2. J. Kakkuri, T. J. Kukkamäki, J.-J. Levallois, and H. Moritz, *Le 250$^e$ anniversaire de la mesure de l'arc du meridien en Lapone* (Helsinki: Suomen Geodeettisen Laitoksen Julkaisuja [Finnish Geodetic Institute], 1986).

3. Smith, *Meridian Arc Measure.*

4. P. Moore and P. Collins, *Astronomy in Southern Africa,* 47.

CHAPTER 6

1. *The Catholic Encyclopedia,* Vol. 2 (New York: Robert Appleton Company).

2. J. B. Salmond, *Wade in Scotland* (Edinburgh; Moray Press, 1934), 21.

3. Ibid., 28–39.

4. George Wade, *General Wade's Report to King George I—1724,* as given in *Highlander Web Magazine,* 2001. Available at http://highlanderweb.co.uk

5. Salmond, *Wade in Scotland,* 52–53.

6. Samuel Johnson, *A Journey to the Western Islands of Scotland* (London: Strahan and Cadell, 1775), 21.

7. Salmond, *Wade in Scotland,* 71.

8. William Taylor, *The Military Roads in Scotland* (London: David & Charles, 1976), 17–45.

9. Samuel Smiles, *The Life of Thomas Telford,* transcribed by Project Gutenberg E-text, 2000.

10. R. Broglio, *Mapping England's Land and Sky: Science Informing Romantic Nationalism,* (Tuscaloosa: University of Alabama, 2001).

11. Ibid.

12. Taylor, *Military Roads in Scotland,* 35.


CHAPTER 7


1. R. M. Gunn, *The Highland Clearances: Its Causes, Effects and Results* (Edinburgh: ScotWeb History Online, 1999). Accessed at Skyelander@Scotlandmail.com

2. William Roy, "An Account of the Measurement of a Base on Hounslow-Heath," *Philosophical Transactions of the Royal Society of London* 75 (1785): 385–440.

3. Yolande O'Donoghue, *William Roy, 1726–1790* (London: British Museum Publications, 1977), 7.

4. Sir Charles Close, *The Early Years of the Ordnance Survey* (London: Institution of Royal Engineers, 1926), 2–4.

5. William Mudge and Isaac Dalby, *An Account of the Operations carried on for accomplishing a Trigonometrical Survey of England and Wales* (London: William Faden of Charing Cross, 1799), 372–392.

6. William Taylor, *The Military Roads in Scotland* (London: Davis & Charles, 1976), 30.

7. O'Donoghue, *William Roy,* 3.

8. Close, *Early Years of the Ordnance Survey,* 2–3.

9. O'Donoghue, *William Roy,* 6.

10. Ibid., 3.

11. Close, *Early Years of the Ordnance Survey,* 3.

12. Mudge and Dalby, *Trigonometrical Survey,* 372.

13. Close, *Early Years of the Ordnance Survey,* 4.

14. Ibid., 3.


CHAPTER 8


1. Nevil Maskelyne, "Concerning the Latitude and Longitude of the Royal Observatory at Greenwich with Remarks on a Memorial of the late M Cassini de Thury," *Philosophical Transactions of the Royal Society of London* 77 (1787): 151–187.

2. James Boswell, *The Life of Samuel Johnson* (London: David Campbell [1791] 1992), 287.

3. Derek Howse, *Nevil Maskelyne: The Seaman's Astronomer* (Cambridge: Cambridge University Press, 1989), 15.

4. Charles Moreland and David Bannister, *Antique Maps—A Collector's Handbook* (London: 2000). Available at http://www.antiquemaps.co.uk/book/index.html

5. Ibid.

6. Hans Woolf, *Transits of Venus: A Study in Eighteenth-Century Science* (Princeton, N.J.: Princeton University Press, 1959).

7. George S. Ritchie, *The Admiralty Chart* (Durham, Eng.: Pentland Press, 1995), 25–28.

8. Brian Connell, *The Plains of Abraham* (London: Hodder & Stoughton, 1960), 161–166.

9. John Keay, *The Honourable Company: A History of the English East India Company* (London: Harper Collins, 1993), 344.

10. Jeremiah Dixon, Letter to the Royal Society, written from Plymouth, January 1763.

11. Charles Mason, "Observations at the Cape of Good Hope," MS Add. 8.1, Royal Astronomical Society, 215.

CHAPTER 9

1. Edwin Danson, *Drawing the Line: How Mason and Dixon Surveyed the Most Famous Border in America* (New York: Wiley, 2000).

2. Charles Mason, Letter to the Gentlemen Commissioners for Dividing the Provinces of Maryland and Pennsylvania, 8 January 1768.

3. Nevil Maskelyne, "Introduction to the following Observations, made by Messieurs Charles Mason and Jeremiah Dixon, for determining the Length of a Degree of Latitude, in the Provinces of Maryland and Pennsylvania, in North America," *Philosophical Transactions of the Royal Society of London* 58 (1769): 273.

4. William Mudge and Isaac Dalby, *An Account of the Operations carried on for accomplishing a Trigonometrical Survey of England and Wales* (London: William Faden of Charing Cross, 1799), 372–392.

5. Sir Charles Close, *The Early Years of the Ordnance Survey* (London: Institution of Royal Engineers, 1926), 5.

6. James Boswell, The *Life of Samuel Johnson* (David Campbell [1791], 1992), 229.

7. H. W. Brands, *The First American: The Life and Times of Benjamin Franklin* (New York: Anchor Books, 2002), 332.

8. *Philosophical Transactions of the Royal Society of London* 59 (1766).

9. Richard Hough, *Captain James Cook: A Biography* (London: Hodder & Stoughton, 1994), 45–46.

10. George S. Ritchie, *The Admiralty Chart* (Durham, The Pentland Press, 1995), 27–28.

11. J. C. Beaglehole, *The Journals of Captain James Cook on His Voyages of Discovery* (Cambridge: Hakluyt Society, Cambridge University Press, 1974).

12. Ibid.

## CHAPTER 10

1. Thomas D Cope, "Charles Mason, Jeremiah Dixon and the Royal Society," *Notes and Records of the Royal Society of London* 9 (October 1951).

2. Royal Society of London, MS 30.

3. Charles Mason, "Journal of papers prepared for the Royal Society on the Transit of Venus observed in Ireland, 1769," MS Add. 8.1, Royal Astronomical Society.

4. *Philosophical Transactions of the Royal Society of London* 59, 262.

5. James Boswell, The *Life of Samuel Johnson* (London: David Campbell [1791], 1992), 771.

6. Thomas Penn, Letter to the American Philosophical Society, in *Philosophical Transactions of the Royal Society of London* 59 (1770).

7. J. C. Beaglehole, *The Journals of Captain James Cook on His Voyages of Discovery* (Cambridge: Hakluyt Society, Cambridge University Press, 1974).

8. Stephanie Pain, "For the love of Venus," *New Scientist* no. 182: June 2004.

9. Mason, *Journal for the Transit of Venus*.

10. Ibid.

11. Boswell, *Life of Johnson*.

12. J. Dixon, and W. Bayly, "Observations Made on the Island of Hammerseft," *Philosophical Transactions of the Royal Society of London* 59 (1770).

13. R. D. Davies, "A Commemoration of Maskelyne at Schiehallion," *Quarterly Journal of the Royal Astronomical Society* 26 (1985) 289–294.

## CHAPTER 11

1. William Roy, Letter to James Lind, 26 October 1771, Colby Collection, Close. 6.

2. *Dictionary of Scientific Biography* (Oxford: Oxford University Press).

3. T. Radford, "Frankenstein May Have Been Based on Scots Scientist," *Manchester Guardian,* 1 May 2002.

4. J. C. Beaglehole, *The Journals of Captain James Cook on His Voyages of Discovery* (Cambridge. Hakluyt Society, Cambridge University Press, 1974).

5. Nevil Maskelyne, "A Proposal for measuring the Attraction of some Hill in this Kingdom by astronomical Observations," *Philosophical Transactions of the Royal Society of London* 65, (1775 [read 1772]): 495–99.

6. Nevil Maskelyne, Letter to John Nourse, Greenwich, 20 January 1767. Courtesy Maggs Bros. Ltd, London, 2001.

7. James Boswell, The *Life of Samuel Johnson* (London: David Campbell [1791], 1992), 524.

8. Maskelyne, "Proposal for measuring."

9. Ibid.

## CHAPTER 12

1. Nevil Maskelyne, Letter to James Lind at Edinburgh, 3 August 1773, MS Ma12, Royal Society of London.

2. Ibid.

3. Ibid.

4. Ibid.

5. C. Phipps, *A Voyage towards the North Pole, undertaken in the year 1773, by His Majesty's Command, for making Discoveries and Nautical and Philosophical Observations in the Northern Seas.* (London: J. Nourse, 1775).

6. Samuel Smiles. *The Life of Thomas Telford, Civil Engineer.* Project Gutenberg e-text, 2000.

7. Ibid.

8. Samuel Johnson, *A Journey to the Western Islands of Scotland* (London: Strahan and Cadell, 1775), 35.

9. Ibid., 33.

10. James Boswell, *The Life of Samuel Johnson* (London: David Campbell [1791], 1992), 268.

11. Johnson, *Journey to the Western Islands*, 35.


CHAPTER 13

1. A. De Morgan, *Notes and Queries,* as quoted in Howse.

2. Nevil Maskelyne, "An Account of Observations made on the Mountain Schehallien for finding its Attraction," *Philosophical Transactions of the Royal Society of London* 65 (1775): 500–542. This is the definitive account of the Schiehallion experiment but is economical in its acknowledgement of the work of others.

3. Ibid., 526.

4. Ibid., 505.

5. Ibid., 512.

6. Reuben Burrow, Letter to Royal Society, MS MM4.81, Royal Society.

7. Richard M. Sillitto, "Maskelyne on Schiehallion," text of lecture delivered to Royal College of Science and Technology, Glasgow, 1957.

8. William Roy, Letter to James Lind, Lanark, 26 July 1774, Colby Collection, Close, 7.

9. Ibid.

10. Samuel Smiles, *The Life of Thomas Telford, Civil Engineer.* Project Gutenberg e-text, 2000.


CHAPTER 14

1. Nevil Maskelyne, "An Account of Observations made on the Mountain Schehallien for finding its Attraction," *Royal Society of London* 65 (1775): 500–542.

2. Nevil Maskelyne, Letter to James Lind: From the Observatory, in the south side of Shihallien, 18 July 1774, MS Ma13, Royal Society.

3. William Roy, Letter to James Lind dated 26 July 1774, Colby Collection, Close, 7.

4. Reuben Burrow, "The Journals of Reuben Burrow," MSS Add. 7.3, Royal Astronomical Society.

5. William Roy, Letter to James Lind, Lanark, 5 September 1774, Colby Collection, Close, 8.

6. Charles Hutton, "An Account of the Calculations made from the Survey and Measures taken at Schehallien, in order to ascertain the mean Density of the Earth," *Philosophical Transactions of the Royal Society of London* 68 (1778): 705–710. Maskelyne fails to mention this fact.

7. Maskelyne, "An Account of Observations," 524–525.

8. Nevil Maskelyne (attributed), as quoted in Derek Howse, *Nevil Maskelyne: The Seaman's Astronomer* (Cambridge: Cambridge University Press, 1989), 138.

## CHAPTER 15

1. Reuben Burrow, Letter to Royal Society, Greenwich 26 January 1775, MS MM4.81, Royal Society.

2. Ibid.

3. Nevil Maskelyne, Letter to Rev. Digby Marsh, 29 November 1790, as quoted in Derek Howse, *Nevil Maskelyne: The Seaman's Astronomer* (Cambridge: Cambridge University Press, 1989), 138–139.

4. H. W. Brands, *The First American* (New York: First Anchor Books, 2002), 269–275.

5. Ibid., 274.

6. Ibid., 190.

7. R. W. Emerson, *"Concord Hymn,"* written for the July 4 celebrations of 1837.

8. *Concise Oxford Dictionery of Words and Phases* (Guild Publishing, 1981), 193 (3).

9. Nevil Maskelyne, "An Account of Observations made on the Mountain Schehallien for finding its Attraction," *Royal Society of London* 65 (1775): 531.

10. Ibid., 534.

11. Ibid., 533.

12. Howse, *Nevil Maskelyne,* 140.

13. *St. James's Chronicle* no. 2327 (January 1776): 11–13, quoted in ibid., 140.

14. William Roy, Letter to James Lind 15 June 1775, quoted in Sir Charles Close, *The Early Years of the Ordnance Survey* (London: Institution of Royal Engineers, 1926), 8.

15. William Roy, Letter to James Lind 13 January 1776, quoted in ibid, 9.

16. James Boswell, *The Life of Samuel Johnson* (London: David Campbell [1791], 1992), 931.

17. Brands, *The First American,* 540–542.

18. Boswell, *Life of Johnson,* 740.

19. Charles Hutton, "An Account of the Calculations from the Survey and Measures taken at Schehallien, in order to ascertain the mean Density of the Earth," *Philosophical Transactions of the Royal Society of London* 68 (1778): 689–788. This paper is the chief source of information on the experimental process.

## CHAPTER 16

1. James Boswell, *The Life of Samuel Johnson* (London: David Campbell [1791], 1992), 458.

2. Ramsey A. Muir, *A Short History of the British Commonwealth,* vol. 1 pt. 3 (London: George Philip, 1930), 773.

3. John Keay, *The Honourable Company: A History of the English East India Company* (London, HarperCollins, 1993), 324.

4. Ibid., 326.

CHAPTER 17

1. Clements R. Markham, *Major James Rennell and the rise of modern English geography* (New York: Macmillan, 1895). Markham's book remains the definitive accout of the life of James Rennell and is used throughout this text.

2. Ibid., 41.

3. Ibid., 44.

4. Ibid., 45.

5. John Keay, *The Great Arc* (London: HarperCollins, 2000), 38–40.

6. Markham, *James Rennell,* 46.

7. Ibid., 61.

8. Ibid., 53.

9. Ibid., 63.

10. W. W. Ristow, *Christopher Colle's Survey of the Roads, 1789* (St. Lawrenceburg, Ind.: Surveyors Historical Society, 2001), 7.

11. Rajesh Kochhar, *Madras and Kodaikanal Observatories: A Brief History* (New Delhi: Resonance, August 2002).

12. John Keay, *The Honourable Company: A History of the English East India Company* (London: HarperCollins, 1993), 407.

CHAPTER 18

1. Clements R. Markham, *Major James Rennell and the rise of modern English geography* (New York: Macmillan, 1895) 96–97.

2. César-François Cassini de Thury, "Sur la junction de Douvres à Londres. Par M. Cassini de Thury, Directeur de l'Observatoire Royal; de la Société Royale de Londres, etc." Paris, 1783.

3. Sir Charles Close, *The Early Years of the Ordnance Survey* (London: Institution of Royal Engineers, 1926), 14.

4. William Mudge and Isaac Dalby, *An Account of the Operations carried on for accomplishing a Trigonometrical Survey of England and Wales* (London: William Faden, 1799), 374.

5. Ibid., 374.

6. Ibid., 375.

7. Sir Charles Close, *The Early Years of the Ordnance Survey,* 23.

8. Ibid., 15.

9. Jesse Ramsden, Letter to Mr Goodriche dated London, 24 August 1784. Reproduced by kind permission of Roy Davids Ltd., Great Haseley, Oxford.

10. William Roy, MS DM 4–23, Royal Society.

11. Charles Blagden, Letter to Joseph Banks, 27 December 1783, as quoted in Derek Howse, *Nevil Maskelyne: The Seaman's Astronomer* (Cambridge: Cambridge University Press, 1989), 159, 255.

12. Derek Howse, *Nevil Maskelyne,* 152–154.

13. Ibid., 152–4.

14. Close, *The Early Years of the Ordnance Survey,* 19.

15. Mudge and Dalby, *Trigonometrical Survey,* 375.

16. Ibid., 378.

17. Ibid., 378.

18. William Roy, Draft of paper to the Royal Society 1790, as quoted in Yolande O'Donoghue, *William Roy, 1726-1790* (London: British Museum Publications, 1977), 49.

### CHAPTER 19

1. Isaac Dalby, "A Short Account of the late Mr Reuben Burrow's Measurement of a Degree of Longitude and Another of Latitude, Under the Tropic in Bengal in the years 1790 and 1791" (fragment of pamphlet) (Calcutta: Asiatick Researches, ca. 1793).

2. Clements R. Markham, *Major James Rennell and the rise of modern English geography* (New York: Macmillan, 1895), 98.

3 Dalby, "Burrow's Measurement."

4. Nevil Maskelyne, Letter to Digby Marsh, 29 November, 1790.

5. Reuben Burrow, *The Journals of Reuben Burrow,* vols. 1 and 2, 1775–88. MS Add. 7.3, London, Royal Astronomical Society.

6. Ibid.

7. Ibid.

8. Ibid.

9. Ibid.

10. Ibid.

11. Ibid.

12. Ibid.

13. Ibid.

14. Reuben Burrow, *A Proof that the Hindoos had the Binomial Theorem,* vol. 2 (Calcutta: Asiatick Researches, 1790).

15. Burrow, *Journals.*

16. M. H. Edney, *Mapping an Empire: The Geographic Construction of British India, 1765-1843* (Chicago: University of Chicago Press, 1997).

17. Clements R. Markham, *Major James Rennell and the rise of modern English geography* (New York, Macmillan, 1895), 98.

18. Burrow, *Journals,* 6 January 1775.

19. Dalby, "Burrow's Measurement."

### CHAPTER 20

1. Isaac Dalby, "A Short Account of the late Mr Reuben Burrow's Measurement of a Degree of Longitude and Another of Latitude under the Tropic in Bengal in the years 1790 and 1791" (Calcutta: Asiatick Researches). This work provides the chronology of events and details of the measurements undertaken by Burrow.

2. R. H. Phillimore, *Historical Records of the Survey of India* (Dehra Dun, India: 1958), 185.

3. Rajesh Kochhar, *Madras and Kodaikanal Observatories: A Brief History* (New Delhi: Resonance, August 2002).

4. Reuben Burrow, Letter to Sir William Jones, quoted in Isaac Dalby, "A Short Account of the late Mr Reuben Burrow's Measurement of a Degree of Longitude and Another of Latitude, Under the Tropic in Bengal in the years 1790 and 1791" (fragment of pamphlet) (Calcutta: Asiatick Researches, ca. 1793).

5. Isaac Dalby, "A Short Account of the late Mr Reuben Burrow's Measurement of a Degree of Longitude and Another of Latitude, Under the Tropic in Bengal in the years 1790 and 1791" (fragment of pamphlet) (Calcutta: Asiatick Researches, ca. 1793).

6. Ibid.

7. Ibid.

8. Reuben Burrow, *The Journals of Reuben Burrow*, vol. 3, 1788, MS Add. 7.3, London, Royal Astronomical Society. Comprising mostly household accounts and sundry calculations at the time of the survey, these writings provide background information and clues to Burrow's state of mind at the time.

9. Dalby, "Burrow's Measurement."

10. Ibid.

11. Ibid.

12. Ibid.

13. Ibid.

14. Ibid.


CHAPTER 21

1. Jesse Ramsden, Letter to Council of the Royal Society, Piccadilly 13 May 1790. MM 3.30, Royal Society.

2. William Roy, "An Account of the Trigonometrical Operation, whereby the Distance between the Meridians of the Royal Observatories of Greenwich and Paris has been determined," *Philosophical Transactions of the Royal Society of London* 80, pt. 1 (1790): 111–270.

3. William Mudge and Isaac Dalby, *An Account of the Operations carried on for accomplishing a Trigonometrical Survey of England and Wales* (London: William Faden, 1799), 372–392.

4. D. Burridge, *The Royal Commission Defences of Dover*, private printing, Dover, 1989, 6.

5. Memoirs of the Mudge Family, quoted in Close, *The Early Years of the Ordnance Survey* (London: Institute of Royal Engineers, 1926), 30.

6. Mudge and Dalby, *Trigonometrical Survey*, 382.

7. Ibid., 382.

8. Ibid., 383.

9. Sir Charles Close, *The Early Years of the Ordnance Survey* (London: Institute of Royal Engineers, 1926) 25, 30–32.

10. Ibid., 43.

11. Alan Stimson, *The Quest for Longitude*. Proceedings of the Longitude Symposium, Harvard University, Cambridge, Mass., Nov. 4–6, 1993, ed. W. J. H. Andrewes (Cambidge, Mass.: Collection of Historical Scientific Instruments, Harvard University, 1995).

12. Ibid.

13. Close, *Ordnance Survey*, 43.

14. Ibid., 37.

15. Ibid., 37.

16. Richard Hough, *James Cook* (London: Hodder & Stoughton, 1995), 220.

17. Mudge and Dalby, *Trigonometrical Survey*, 387.

18. Ibid., 384.

19. Ibid., 384–385.

20. Ibid., 388.

CHAPTER 22

1. Thomas Glattard and Alfred Bollinger, *Swiss Precision for US Mapping: Ferdinand Rudolf Hassler*. Surveying and Mapping the Americas, FIG, Washington 2002 (Frederick, Md.: Proceedings of the International Institution for History of Surveying & Measurement, Professional Surveyors Publishing Co., 2002).

2. F. Cajori, *The Chequered Career of Ferdinand Rudolph Hassler* (Boston: Chesset Press, 1929).

3. Albert E. Theberge, *The Coast Survey 1807–1867: Vol. 1, History of the Commissioned Corps of the National Oceanic and Atmospheric Administration. Part 1—The Hassler Legacy* (Washington, D.C., National Oceanic and Atmospheric Administration, 2001).

4. Glattard and Bollinger, *Swiss Precision for US Mapping.*

5. Theberge, *Coast Survey.*

6. Ibid.

7. Ibid.

8. R. S. Preston, "The Accuracy of the Astronomical Observations of Lewis and Clark," *Proceedings of the American Philosophical Society* 144, no. 2 (June 2000): 168–191.

9. F. R. Hassler, "Papers on Various Subjects connected with the Survey of the Coast of the United States," *Transactions of the American Philosophical Society* 2, New Series (1825): 232.

10. Glattard and Bollinger, *Swiss Precision for US Mapping.*

11. Preston, *Astronomical observations of Lewis and Clark,* 168–191.

CHAPTER 23

1. F. G. Skinner, *Weights and Measures. Their Ancient Origins and Development in Great Britain* (London, HMSO, 1967).

2. Ibid.

3. Charles Maurice de Talleyrand Perigord, bishop of Autun. Letter to Sir John Riggs Miller 29 March 1790.

4. Pierre Aubert, "Borda et le système métrique," *Bulletin de la Société Borda*, no. 456, 2000.

5. Ibid.

6. Ibid.

7. Ibid.

8. C. Wolf, "Recherches historiques sur les étalons de l'Observatoire," and Suzanne Débarbat, "Les Règles de Borda: A Secondary Standard in Geodesy," both from *Annals of Chemistry & Physics,* Series 5 (2000).

9. Aubert, "Borda et le système métrique."

## CHAPTER 24

1. William Mudge and Isaac Dalby, *An Account of the Operations carried on for accomplishing a Trigonometrical Survey of England and Wales* (London, William Faden, 1799), 390.

2. Ibid., 390.

3. Sir Charles Close, *The Early Years of the Ordnance Survey* (London: Institute of Royal Engineers, 1926), 50.

4. Ibid., 68.

5. Derek Howse, *Nevil Maskelyne: The Seaman's Astonomer* (Cambridge: Cambridge University Press, 1989), 130.

6. Henry Cavendish, "Experiments to determine the Density of the Earth," *Philosophical Transactions of the Royal Society of London,* 1798.

7. Board of Longitude papers vol. 11, 1802, London: Royal Observatory Greenwich.

8. Sir George Everest, Letter to Thomas Colby, 27 June 1829, Oriental Club, Hanover Square. Colby Collection.

9. Close, *Early Years of Ordnance Survey.*

# Bibliography

Andrewes, W. J. H., ed. *The Quest for Longitude* Proceedings of the Longitude Symposium, Harvard University, Cambridge, Mass., Nov 4–6, 1993. Cambidge, Mass.:
Collection of Historical Scientific Instruments, Harvard University, 1995.

Beaglehole, J. C. *The Journals of Captain James Cook on his Voyages of Discovery.* Cambridge: Hakluyt Society, Cambridge University Press, 1974.

Boswell, James. *The Life of Samuel Johnson,* London: David Campbell, [1791] 1992.

Brands, H. W. *The First American: The Life and Times of Benjamin Franklin.* New York:
Anchor Books, 2002.

Broglio, R. *Mapping England's Land and Sky: Science Informing Romantic Nationalism.*
Tuscaloosa: University of Alabama, 2001.

Burridge, D. *The Royal Commission Defences of Dover.* Dover, Eng.: Private printing,
1989.

Close, Sir Charles. *The Early Years of the Ordnance Survey.* London: Institution of Royal
Engineers, 1926.

Cohen, J. M., trans. *Christopher Columbus: The Four Voyages.* London: Penguin Books,
1969.

Connell, Brian. *The Plains of Abraham.* London: Hodder and Stoughton, 1960.

Crane, Nicholas. *Mercator: The Man Who Mapped the Planet.* London: Weidenfeld &
Nicolson, 2002.

Danson, Edwin. *Drawing the Line: How Mason and Dixon Surveyed the Most Famous Border in America.* New York: Wiley, 2000.

Edney, M. H. *Mapping an Empire: The Geographic Construction of British India, 1765–1843.*
Chicago: University of Chicago Press, 1997.

Engels, Frederich. *Herrn Eugen Dührings Umwälzung der Wissenschaft.* Berlin: 1948.

Gorst, Martin. *Measuring Eternity.* New York: Broadway Books, 2001.

Hough, Richard. *Captain James Cook: A Biography.* London: Hodder & Stoughton, 1994.

Howse, Derek. *Nevil Maskelyne: The Seaman's Astronomer.* Cambridge: Cambridge University Press, 1989.

Johnson, Samuel. *A Journey to the Western Islands of Scotland.* London: Strahan and
Cadell, 1775.

Keay, John. *The Honourable Company: A History of the English East India Company.* London: Harper Collins, 1993.

Keay, John. *The Great Arc.* London, Harper Collins, 2000.

Markham, Clements R. *Major James Rennell and the rise of modern English geography.* New York: Macmillan, 1895.

Moreland, Charles, and David Bannister. *Antique Maps—A Collector's Handbook.* London: Web edition 2000.

Mudge, William, and Isaac Dalby. *An Account of the Operations carried on for accomplishing a Trigonometrical Survey of England and Wales; begun under the Direction of the Royal Society, and continued by order of the Honourable Board of Ordnance.* London: William Faden of Charing Cross, 1799.

Muir, Ramsey A. *A Short History of the British Commonwealth,* vol. 1 pt. 3. London: George Philip, 1930.

O'Donoghue, Yolande. *William Roy, 1726–1790.* London: British Museum Publications, 1977.

Phillimore, R. H. *Historical Records of the Survey of India,* 4 vols. Dehra Dun, India, 1958.

Ritchie, George S. *The Admiralty Chart.* Durham, Eng.: Pentland Press, 1995.

Salmond, J. B. *Wade in Scotland.* Edinburgh: Moray Press, 1934.

Smiles, Samuel. *The Life of Thomas Telford.* Transcribed by Project Gutenberg E-text, 2000.

Taylor, William. *The Military Roads in Scotland.* London: David & Charles, 1976.

Wolf, C. "Recherches historiques sur les étalons de l'Observatoire." Débarbat, Suzanne, "Les Règles de Borda: A Secondary Standard in Geodesy." Both from *Annals of Chemistry & Physics,* Series 5 (2000).

Woolf, Hans. *Transits of Venus: A Study in Eighteenth-Century Science.* Princeton, N.J.: Princeton University Press, 1959.

JOURNALS

Burrow, Reuben. *The Journals of Reuben Burrow,* vols. 1 and 2, 1775–1788. MS Add. 7.3, Royal Astronomical Society.

Mason, Charles. "Observations at the Cape of Good Hope," MS Add. 8.1, Royal Astronomical Society, 1770.

Mason, Charles. "Journal of papers prepared for the Royal Society on the Transit of Venus observed in Ireland, 1769." MS Add. 8.1, Royal Astronomical Society, 1770.

LETTERS

Blagden, Charles. Letter to Joseph Banks, 27 December 1783.

Burrow, Reuben. Letter to Royal Society. MS MM4.81, Royal Society of London, 1775.

Burrow, Reuben. Letter to Royal Society, Greenwich, 26 January 1775. MS MM4.81, Royal Society of London.

Burrow, Reuben. Letter to Sir William Jones. Quoted in Isaac Dalby, "A Short Account of the late Mr Reuben Burrow's Measurement of a Degree of Longitude and Another of Latitude, Under the Tropic in Bengal in the years 1790 and 1791" (fragment of pamphlet). Calcutta: Asiatick Researches, ca. 1793.

Dixon, Jeremiah. Letter to the Royal Society, written from Plymouth, January 1763.

Everest, Sir George. Letter to Thomas Colby, 27 June 1829, Oriental Club, Hanover Square. Colby Collection.

Maskelyne, Nevil. Letter to John Nourse, Greenwich 20 January 1767. Courtesy Maggs Bros. Ltd, London, 2001.

Maskelyne, Nevil. Letter to James Lind at Edinburgh, 3 August 1773. MS Ma12, Royal Society of London.

Maskelyne, Nevil. Letter to James Lind: From the Observatory, in the south side of Shihallien, 18 July 1774. MS Ma13, Royal Society of London.

Maskelyne, Nevil. Letter to Rev Digby Marsh, 29 November 1790.

Mason, Charles. Letter to the Gentlemen Commissioners for Dividing the Provinces of Maryland and Pennsylvania, 8 January 1768.

Penn, Thomas. Letter to the American Philosophical Society. In *Philosophical Transactions of the Royal Society of London* 59 (1770).

Ramsden, Jesse. Letter to Mr. Goodriche, London, 24 August 1784. Reproduced by kind permission of Roy Davids Ltd, Great Haseley, Oxford.

Ramsden, Jesse. Letter to Council of the Royal Society, Piccadilly 13 May 1790. MM 3.30, Royal Society of London.

Roy, William. Letter to James Lind, 26 October 1771.

Roy, William. Letter to James Lind, Lanark, 26 July 1774. Colby Collection, Close, 7.

Roy, William. Letter to James Lind, Lanark 5 September 1774. Colby Collection, Close, 8.

Roy, William. Letter to James Lind, 15 June 1775.

Roy, William. Letter to James Lind, 13 January 1776.

Roy, William. MS DM 4–23, Royal Society of London.

Tallyrand, Charles Maurice de. Letter to Sir John Riggs, 29 March 1790.

PAPERS AND ARTICLES

Aubert, Pierre. "Borda et le système métrique." *Bulletin de la Société Borda,* no. 456 (2000).

Burrow, Reuben. *A Proof that the Hindoos had the Binomial Theorem,* vol. 2. Calcutta: Asiatick Researches, 1790.

Cassini de Thury, César-François. "Sur la junction de Douvres à Londres. Par M. Cassini de Thury, Directeur de l'Observatoire Royal; de la Société Royale de Londres, etc." Paris, 1783.

Cope, Thomas D. "Charles Mason, Jeremiah Dixon and the Royal Society." *Notes and Records of the Royal Society of London* 9 (October 1951).

Costabel, P and M. Martinet. "Morin, Quelques savants et amateurs de science au XVIIe siècle." Paris, 1986.

Dalby, Isaac. "A Short Account of the late Mr Reuben Burrow's Measurement of a Degree of Longitude and Another of Latitude, Under the Tropic in Bengal in the years 1790 and 1791." Fragment of pamphlet. Calcutta: Asiatick Researches, ca. 1793.

Danson, Edwin. *Jeremiah's Theodolite.* London: Surveying World, 2001.

Davies, R. D. "A Commemoration of Maskelyne at Schiehallion." *Quarterly Journal of the Royal Astronomical Society* 26 (1985).

Dixon, J., and W Bayley. "Observations made on the island of Hammerfest." *Philosophical Transactions of the Royal Society of London* 59 (1770).

Hassler, F. R. "Papers on Various Subjects connected with the Survey of the Coast of the United States." *Transactions of the American Philosophical Society* 2, New Series (1825).

Hutton, Charles. "An Account of the Calculations made from the Survey and Measures taken at Schehallien, in order to ascertain the mean Density of the earth." *Philosophical Transactions of the Royal Society of London* 68 (1778).

Kakkuri, J., T. J. Kukkamäki, J.-J. Levallois, and H. Moritz. *Le 250ᵉ anniversaire de la mesure de l'arc du meridien en Lapone.* Helsinki: Suomen Geodeettisen Laitoksen Julkaisuja (Finnish Geodetic Institute), 1986.

Kochhar, Rajesh. *Madras and Kodaikanal Observatories: A Brief History.* New Delhi: Resonance, August 2002.

Lelgemann, Dieter. *On the Ancient Determination of the Meridian Length by Eratosthenes of Kyrene.* Presentation at History of Surveying and Mapping WSHS1, FIG Working Week, Athens, 2001.

Marguet, F. "Histoire de la longitude à la mer au XVIIIc siècle." Paris, 1917.

Maskelyne, Nevil. "Concerning the Latitude and Longitude of the Royal Observatory at Greenwich with Remarks on a Memorial of the late M Cassini de Thury." *Philosophical Transactions of the Royal Society of London* 77 (1787).

Maskelyne, Nevil. "Introduction to the following Observations, made by Messieurs Charles Mason and Jeremiah Dixon, for determining the Length of a Degree of Latitude, in the Provinces of Maryland and Pennsylvania, in North America." *Philosophical Transactions of the Royal Society of London* 58, (1769).

Maskelyne, Nevil. "A Proposal for measuring the Attraction of some Hill in this Kingdom by astronomical Observations." *Philosophical Transactions of the Royal Society of London* 65 (1775).

Maskelyne, Nevil. "An Account of Observations made on the Mountain Schehallien for finding its Attraction." *Philosophical Transactions of the Royal Society of London* 65 (1775).

Maupertuis, Pierre de. *Relation du voyage fait par ordre du Roi au cercle polaire pour déterminer la figure de la terre.* Paris: Mémoires de l'Academie Royale des Sciences, 1737.

Moren, Gonzalez Antonio, and Pesar la Tierra. *Test y Origen de un Anacronismo.* Madrid: Departamento de Didáctica de las Ciencias Experimentales, 2000.

Pain, Stephanie. "For the love of Venus." *New Scientist* no. 182 (June 2004).

Phipps, C. *A Voyage towards the North Pole, undertaken in the year 1773, by His Majesty's Command, for making Discoveries and Nautical and Philosophical Observations in the Northern Seas.* London: J. Nourse, 1775.

Preston, R. S. "The accuracy of the astronomical observations of Lewis and Clark." *Proceedings of the American Philosophical Society* 144, no. 2 (June 2000): 168, 191.

Radford, T. "Frankenstein May Have Been Based on Scots Scientist." *Manchester Guardian,* 1 May, 2002.

Ristow, W. W. *Christopher Colle's Survey of the Roads, 1789.* St Lawrenceburg, Ind.: Surveyors Historical Society, 2001.

Roy, William. "An Account of the Measurement of a Base on Hounslow-Heath." *Philosophical Transactions of the Royal Society of London* 75 (1785).

Roy, William. "An Account of the Trigonometrical Operation, whereby the Distance between the Meridians of the Royal Observatories of Greenwich and Paris has been determined." *Philosophical Transactions of the Royal Society of London* 80, pt. 1 (1790).

Sillitto, Richard M. "Maskelyne on Schiehallion." Text of lecture delivered to Royal College of Science and Technology, Glasgow, 1957.

Smith, James R. *The Meridian Arc Measure in Peru, 1735-1745. Surveying and Mapping the Americas.* Washington, D.C.: Federation International de Geometre Convention, 2002.

## WEB SITES

Gunn, R. M., *The Highland Clearances: Its Causes, Effects and Results.* Edinburgh: Scot-Web History Online, 1999. Accessed at Skyelander@Scotlandmail.com.

Wade, George. *General Wade's Report to King George I—1724,* as given in *Highlander Web* magazine, 2001. Available at http://highlanderweb.co.uk

## SECONDARY SOURCES

*American National Biography Online.* Available at http://anb.org (American Council of Learned Societies and Oxford University Press).

Amod, A. "A History of Geodetic Surveying in South Africa: The Cape Odyssey." *Cape Town: Historical Media* 2, issue 7, Aug/Sept, 2002.

Bedini, Silvio. "The Transit in the Tower: English Astronomical Instruments in Colonial America." *Annals of Science* 54 (1997).

Besant, Walter. *Captain Cook.* London and New York: Macmillan, 1890.

Burkard, R. K. *Geodesy for the Layman.* Washington, D.C.: U.S. Defense Mapping Agency [1959], 4th ed. 1984.

Cassini, Legendre, and Méchain. *Exposé des Opérations faites en France.* Paris: Academie des Sciences, 1791.

*Catholic Encyclopedia,* vol. 2. Online Edition. Copyright Kevin Knight, Imprimatur, and John M. Farley, archbishop of New York.

*Concise Oxford Dictionary of Quotations,* vol. 1, 2nd ed. Oxford: Oxford University Press, 1981.

Connor, R. D. *The Weights and Measures of England.* London: HMSO, 1978.

Darwin, Charles. *Journal of a Voyage Round the World.* London: Nelson, 1890.

*Dictionary of Scientific Biography.* Oxford: Oxford University Press.

Dixon, Jeremiah. "Observations made on the island of Hammerfest." *Philosophical Transactions of the Royal Society of London* 59 (1769).

Dracup, J. F. "A Brief History of Geodetic Surveys in the United States 1807-2000." Series of articles in ACSM Bulletins (March/April, July/August, September/October 1995).

Fernie, J. D. "Transits, Travels and Tribulations." *American Scientist* nos. 85–87 (1997).

Furneaux, R. *The Seven Years War.* London: Granada, 1973.

Glatthard, T., and A. Bollinger. *Swiss Precision for U.S. Mapping—Ferdinand Rudolph Hassler—First Chief of U.S. Coast and Geodetic Survey and U.S. Bureau of Standards.* Washington, D.C.: Proceedings of the Federation International de Geometre XXII International Congress, 19–26 April 2002.

Hibbert, C. *The English*. Glasgow: William Collins, 1987.

Hughes, I. "The Reverend Neville Maskelyne, Schiehallion and the Force of Gravity." *Edinburgh Journal & News of the John Muir Trust*, Summer 2000.

*Instruction sur les mesures déduites de la grandeur de la Terre, uniforme pour toute la République,et sur les calculs relatifs à leur division decimal*. Paris: Imprimerie Nationale, An II de la République, 1793–1794.

Kippis, A. *A Narrative of the Voyages Round the World performed by Captain James Cook*. London: Bickers, 1893.

Kuczynski, J. *The Rise of the Working Class*. Trans. by C. T. A. Ray. London: World University Library, Weidenfeld & Nicolson, 1967.

Leadstone, G. S. *Maskelyne's Schehallien experiment of 1774*. London: Physics Education, Institute of Physics, 1978.

Lee, C. *This Sceptered Isle*. London and New York: Penguin Books, 1998.

Markham, Clements R. *A Memoir of the Indian Surveys*. London: W. H. Allen, 1872.

Marshall, D. *Eighteenth Century England*. London: Longmans, Green, 1968.

Mason, Charles, and Jeremiah Dixon. *The Journal of Charles Mason and Jeremiah Dixon*. Transcribed from the original by A. Hughlett Mason. Philadelphia: American Philosophical Society, 1969.

"Mason and Dixon at the Cape." Monthly Notes of the Astronomical Society of South Africa, November 1951.

Méchain, Pierre François André. *Backsights Magazine*. Surveyors Historical Society.

*Mesure de Dégré du Meridien en France*. Paris: Mémoires de l'Academie Royale des Sciences, 1736.

*National Dictionary of Biography*. Oxford: Oxford University Press.

Parrott, Sir Edward. *Allies, Foes, and Neutrals*. New York: Thomas Nelson, 1918.

Picard, Lisa. *Dr Johnson's London*. London: Weidenfeld & Nicolson, 2000.

Robertson, John. *A Treatise of Mathematical Instruments*. London: J. Nourse, 1775. Reprint edition with notes by David Manthey. Arlington, Va.: Flower-de-Luce Books, 2002.

Rudé, George. *The French Revolution*. London: Weidenfeld & Nicolson, 1988.

Rule, J. *The Vital Century*. London & New York: Longman, 1992.

Shepherd, William R. *The Historical Atlas*. London: 1923.

Skinner, F. G. *Weights and Measures. Their ancient origins and development in Great Britain up to 1855*. London: HMSO, 1967.

"Two Astronomers in Search of the Metre." *The Economist*, 8 July, 1989.

U.S. National Oceanographic and Atmospheric Agency. Library and information service available at http://noaa.gov

van Geer, Pieter. *Notice sur la vie et les travaux de Willebrord Snellius*. Amsterdam: Archives Neerlandaises de Sciences Exactes et Naturelles, 1883.

Wraight, A. J., and E. B. Roberts. *The Coast and Geodetic Survey 1807–1957*. Washington, D.C.: U.S. Government Printing Office, 1957.

Wilkie, T. "*Time to Re-measure the Metre.*" *New Scientist*, 27 October 1983.

# Index